# UNDERSTANDING THE [

*hat is the unknown? What are the data? What is the*
it possible to satisfy the condition? Is the condition sufficient to
:termine the unknown? Or is it insufficient? Or redundant? Or
•ntradictory?
raw a figure. Introduce suitable notation.
:parate the various parts of the condition. Can you write them
•wn?

## DEVISING A PLAN

ave you seen it before? Or have you seen the same problem in a
ightly different form?
*o you know a related problem?* Do you know a theorem that
•uld be useful?
*•ok at the unknown!* And try to think of a familiar problem having
.e same or a similar unknown.
*ere is a problem related to yours and solved before. Could you
;e it?* Could you use its result? Could you use its method? Should
•u introduce some auxiliary element in order to make its use
•ssible?
ould you restate the problem? Could you restate it still differ-
.tly? Go back to definitions.
you cannot solve the proposed problem try to solve first some
:lated problem. Could you imagine a more accessible related
roblem? A more general problem? A more special problem? An
.alogous problem? Could you solve a part of the problem? Keep
.nly a part of the condition, drop the other part; how far is the
.nknown then determined, how can it vary? Could you derive
•mething useful from the data? Could you think of other data
ppropriate to determine the unknown? Could you change the
.nknown or the data, or both if necessary, so that the new unknown
.nd the new data are nearer to each other?
·id you use all the data? Did you use the whole condition? Have
•u taken into account all essential notions involved in the prob-
·m?

## CARRYING OUT THE PLAN

arrying out your plan of the solution *check each step.* Can you
·ee clearly that the step is correct? Can you prove that it is correct?

## LOOKING BACK

an you *check the result?* Can you check the argument?
an you derive the result differently? Can you see it at a glance?
:an you use the result, or the method, for some other problem?

*Problem Solving*
*in*
*School*
*Mathematics*

# Problem Solving in School Mathematics

## 1980 Yearbook

### Stephen Krulik
1980 Yearbook Editor
Temple University

### Robert E. Reys
General Yearbook Editor
University of Missouri

## National Council of
## Teachers of Mathematics

**Library of Congress Cataloging in Publication Data:**

Main entry under title:

Problem solving in school mathematics.
   (Yearbook—National Council of Teachers of
Mathematics  ; 1980)
      Bibliography: p.
      1. Problem solving. I. Krulik, Stephen.
II. Series: National Council of Teachers of Mathematics.
Yearbook  ; 1980.
QA1.N3  1980  [QA63]  510s [153.4'3]  79-27145
ISBN 0-87353-162-0

*Printed in the United States of America*

# Table of Contents

*Thomas Butts,* Case Western Reserve University, Cleveland, Ohio

Mathematics is problem solving. A primary source of motivation for solving a problem is the manner in which that problem is posed. Different types of mathematical problems are discussed, and many suggestions for improving the art of posing problems are offered.

*Marilyn N. Suydam,* Ohio State University, Columbus, Ohio

The results of research on problems solving from the past decade are untangled, not merely to describe findings but also to call attention to directions for classroom practice that may be indicated. Research about children as problem solvers, about problems, and about problem-solving strategies are presented, with "clues for teaching" highlighted.

*Alan Osborne,* Ohio State University, Columbus, Ohio
*Margaret B. Kasten,* Ohio State University, Columbus, Ohio

Opinions of professional and lay samples about the curriculum for school mathematics in the 1980s were surveyed by the NCTM PRISM Project. Summary observations of the data about the preferred content and approaches to problem solving indicate support but little preference for extremes in methodology or resources.

*Joel Schneider,* Comprehensive School Mathematics Program, CEMREL, Saint Louis, Missouri
*Kevin Saunders,* Northern Michigan University, Marquette, Michigan

One approach to early teaching of problem solving is to provide a pictorial language with which to record information. The availability of such a tool not only encourages good habits but also allows individual freedom in approaching problems. As students progress, pictorial languages may be an effective support in the gradual introduction of symbolic languages. Two examples of pictorial languages illustrate their application.

*Linda J. DeGuire,* University of Georgia, Athens, Georgia

Two classroom episodes are used to attempt to illustrate Polya's style of teaching problem solving. The episodes are set in middle school classrooms, one in fifth grade and one in eighth grade.

*Edward J. Davis,* University of Georgia, Athens, Georgia
*William D. McKillip,* University of Georgia, Athens, Georgia

Story problems are important because they are a major vehicle through which we teach applications. Difficulties with story problems have caused teachers and students to have negative attitudes toward them. This essay addresses ways to overcome such attitudes and to improve students' performance.

*Jeffrey C. Barnett,* Fort Hays State University, Hays, Kansas
*Larry Sowder,* Northern Illinois University, DeKalb, Illinois
*Kenneth E. Vos,* College of Saint Catherine, Saint Paul, Minnesota

> What can a teacher do in teaching story problems? Seek interesting contexts, let the learners make up the problems, present the stories in pictures or with objects. And, pay particular attention to reading factors that are especially important in mathematics—the precise technical vocabulary and the slower, more attentive readings necessary—with emphasis on keys to the mathematical structure of the problem.

*John F. LeBlanc,* Indiana University, Bloomington, Indiana
*Linda Proudfit,* Valparaiso University, Valparaiso, Indiana
*Ian J. Putt,* Townsville College of Advanced Education, Queensland, Australia

> This essay presents an instruction technique for teaching problem solving that is based on Polya's four-stage model of problem solving. This technique emphasizes the teacher's role, particularly during the first stage (understanding) and the final stage (reflecting on the solution). Examples dealing with a standard textbook problem and a "process" problem depict the suggested technique in classroom terms.

*Joan Duea,* Price Laboratory School, University of Northern Iowa, Cedar Falls, Iowa
*George Immerzeel,* Price Laboratory School, University of Northern Iowa, Cedar Falls, Iowa
*Earl Ockenga,* Price Laboratory School, University of Northern Iowa, Cedar Falls, Iowa
*John Tarr,* Price Laboratory School, University of Northern Iowa, Cedar Falls, Iowa

> There are advantages in using calculators in problem solving: (1) instruction can focus on problem-solving processes; (2) guess-and-test using a calculator code and using tables become more meaningful strategies; (3) real-world problems are appropriate in classrooms; and (4) students are more involved through experimentation and data collection. The need for materials to implement calculators in classroom problem solving can be met by modifying textbook problems, developing decks of problem cards, and building a problem file. All are discussed in this paper.

*Marilyn H. Jacobson,* Montgomery County Community Schools Corporation, Bloomington, Indiana
*Frank K. Lester, Jr.,* Indiana University, Bloomington, Indiana
*Arthur Stengel,* Indiana University, Bloomington, Indiana

> A way to get intermediate grade students actively involved in solving process-oriented problems is described in this essay. Discussions of several process problems are included.

John A. *Malone,* Western Australian Institute of Technology, Bentley,
Western Australia
Graham A. *Douglas,* University of Western Australia, Nedlands,
Western Australia
Barry V. *Kissane,* University of Western Australia, Nedlands,
Western Australia
Roland S. *Mortlock,* University of Western Australia, Nedlands,
Western Australia

> The development of tests of problem-solving ability that use a measurement approach based on the work of Georg Rasch (1960) and that overcome some of the traditional difficulties is discussed. A rationale for the use of the Rasch model is presented; the model's characteristics are described along with an outline of a procedure for the compilation of a problem pool and its subsequent use in the construction of problem-solving tests.

Harold L. *Schoen,* University of Iowa, Iowa City, Iowa
Theresa *Oehmke,* University of Iowa, Iowa City, Iowa

> The Iowa Problem Solving Project Test, a paper-and-pencil test that measures a person's ability to understand a verbal problem, to carry out certain problem-solving strategies, and to vary conditions in a problem that has been solved, is described. The test is designed for grades 5–8.

Sarah F. *Mason,* University of Georgia, Athens, Georgia

> A sampling of materials on problem solving that are of primary interest to teachers at all levels. The entries have been organized into six sections: (1) Bibliographies; (2) General Works; (3) Suggestions for Teachers; (4) Puzzles and Recreations; (5) Mathematical Discussions; and (6) Collections of Problems.

# *Foreword*

In 1973 the Publications Committee of the National Council of Teachers of Mathematics initiated a new concept in the development and publication of the Council's yearbooks. It recommended that a yearbook addressing a theme of timely interest to the wide range of NCTM membership be prepared on a schedule that would allow a new one to be presented at each annual meeting. I was asked to serve as the first general editor of this series, and the first of the new yearbooks was published in 1976.

This is the fifth—the final one for which I share responsibility. During this time I have had the pleasure of working closely with the Publications Committee, and I want to express my thanks for their continuous support and cooperation. It is impossible to imagine any committee that is harder working and more dedicated to their task. Specifically, I want to recognize the chairpersons of the Publications Committee with whom I have had the opportunity to work. First, Seaton Smith—who explained the concept but never told me what I was in for—then Jeremy Kilpatrick, George Cathcart, Linda Silvey, Margaret Kenney, and Kay Nebel. Each has been instrumental in shaping and establishing this yearbook series.

As general editor I have had the pleasure of meeting and working with many fine people. Whatever measure of success these yearbooks have achieved is the result of the issue editors, essay authors, reviewers, and the professional staff of NCTM.

Five extraordinary individuals have served as issue editors:

Doyal Nelson, *Measurement in School Mathematics*

F. Joe Crosswhite, *Organizing for Mathematics Instruction*

Marilyn N. Suydam, *Developing Computational Skills*

Sidney Sharron, *Applications in School Mathematics*

Stephen Krulik, *Problem Solving in School Mathematics*

Each of them worked so hard and did his or her job so well that it often made my role perfunctory. My thanks to each of them.

At the heart of each yearbook are the essays and the ideas that are presented. Without exception, the authors of the essays worked hard to produce papers of high quality, adjusted personal schedules to meet production deadlines, and were consistently cooperative in accepting critiques and rewriting to reflect suggestions that were offered. Without their expertise and willingness to contribute their manuscripts, these yearbooks would never have become a reality.

The reviewers are truly the unsung heros, as each editor will readily testify. During the development of these five yearbooks, nearly two hundred different people have contributed their time and talents in this reviewing process. A special thanks to them. It is not possible to name each of them here, but it is lack of space, not gratitude, that prevents this individual recognition.

In addition to the regular editors, there are three people from whom I have consistently sought advice and suggestions during this period. A special thanks to Barbara Bestgen, Douglas Grouws, and Paul Rahmoeller for allowing me to lean on them.

Once all final drafts of the essays have been completed, the professional staff of the NCTM takes over. Their efforts to improve manuscripts and develop them to their full potential is reflected in the final quality of the yearbooks. My thanks to the staff.

I am glad to have had the opportunity to serve as general editor during this inaugural period. It is my hope that this yearbook concept is now firmly established and that it will continue to grow in stature.

ROBERT E. REYS
*General Editor*

# *Preface*

A preface is rarely read by every reader. Many regard it as an unnecessary part of the book, claiming that it just fills up space that might otherwise be put to better use. But this is not always true! A preface is important to a book; it sets the tone, and it gives the author or editor an opportunity to provide readers with some insights—a chance to see just how the book they are about to read came about.

A yearbook of this scope and magnitude doesn't just happen. Actually, *Problem Solving in School Mathematics* represents the work of many Council members over a period of more than three years. In 1977 when the process began, more than fifty mathematics educators from all over the country met in Cincinnati to help develop guidelines for the content and direction of this yearbook. The depth and range of ideas that came out of that meeting literally boggled the mind.

Once the guidelines were established and announced, we received more than eighty drafts for proposed essays on various aspects of problem solving. Each proposal was read by a *minimum* of five qualified reviewers. The papers were then read again by an advisory group and by the editors. It was a difficult task limiting the selection to the twenty-two papers that were finally included. Although space restrictions made us omit many excellent papers, we made a conscientious effort to select those with the widest appeal to potential readers.

From these beginning stages right through to the excellent job done by the Reston staff (who did so much of the final work in actually getting the book into print), literally hundreds of Council members were involved. To each of you who participated, I want to express my personal thanks. Sheer numbers prevent my naming everyone who helped. In particular, however, I should like to thank six people who were involved with this yearbook from start to finish:

| | |
|---|---|
| Jeremy Kilpatrick | University of Georgia, Athens, Georgia |
| Richard Lesh | Northwestern University, Evanston, Illinois |
| Adelyn Muller | Shawnee Mission School District, Shawnee Mission, Kansas |
| Robert Reys | University of Missouri, Columbia, Missouri |
| Jesse Rudnick | Temple University, Philadelphia, Pennsylvania |
| Mark Spikell | George Mason University, Fairfax, Virginia |

Each of the last few decades has been highlighted by a phrase, a slogan, or a major thrust applicable to mathematics education. In the late fifties we had the "revolution in mathematics"; in the sixties we had "modern math"; in the seventies it was "back to basics." As we enter the eighties, the theme appears to be "problem solving." But problem solving is much more than just a phrase. To many people, problem solving is *the* reason for teaching mathematics. It is one of the basic skills that students must take along with them throughout their lives and use long after they have left school. This skill—its teaching and learning—causes many an anxious moment for students as well as for teachers, but it is a skill that can be taught and must be taught!

This book contains something to help every teacher of mathematics teach problem solving. Inside the front and back covers, where they can be easily referred to, are the well-known heuristics of George Polya, the master teacher of problem solving. There is a paper by Polya, papers about Polya. There are papers about research and evaluation. There are papers from Australia, Georgia, California, and many places in between.

Although *Problem Solving in School Mathematics* was never intended to be a definitive work on problem solving (in keeping with the concept of this series of yearbooks), the papers were all written with an eye to the classroom teacher. Here are ideas to be used in the classroom at all levels of instruction. Here are problems, examples, and illustrations. And if these ideas are used, then the authors will have done their job well! Problem solving should be the major focus of all mathematics instruction; *Problem Solving* is an attempt to help the classroom teacher in this important effort.

Glance through the Table of Contents. Pick out an article that interests you. Then jump right in!

STEPHEN KRULIK
*1980 Yearbook Editor*

# On Solving Mathematical Problems in High School

## G. Polya

EDITOR'S NOTE: *This article, although originally presented in the November 1949 issue of the* California Mathematics Council Bulletin *(vol. 7, no. 2), offers some thoughts about problem solving that are as current today as they must have been avant-garde then. It should be read by all teachers of mathematics, not just those who are teaching high school mathematics.*

**1.** SOLVING a problem is finding the unknown means to a distinctly conceived end. If the end by its simple presence does not instantaneously suggest the means, if, therefore, we have to search for the means, reflecting consciously how to attain the end, we have to solve a problem. To solve a problem is to find a way where no way is known off-hand, to find a way out of a difficulty, to find a way around an obstacle, to attain a desired end, that is not immediately attainable, by appropriate means.

Solving problems is the specific achievement of intelligence, and intelligence is the specific gift of man. The ability to go round an obstacle, to undertake an indirect course where no direct course presents itself, raises the clever animal above the dull one, raises man far above the most clever animals, and men of talent above their fellow men.

Solving problems is human nature itself. We may characterize man as the "problem-solving animal"; his days are filled with not immediately attainable aspirations. The greater part of our conscious thinking is concerned with problems; when we do not indulge in mere musing or daydreaming, our thoughts are directed towards some end.

2. If education fails to contribute to the development of the intelligence, it is obviously incomplete. Yet intelligence is essentially the ability to solve problems: every day problems, personal problems, social problems, scientific problems, puzzles, all

Dr. Polya emphasizes that much of student interest and motivation must stem from mathematics itself, from certain intrinsic qualities inherent in mathematics and in the problem solving process. This article indicates how a teacher having an appreciation of this point of view may teach mathematics in such a way that "the student may experience the tension and enjoy the triumph of discovery."

Dr. Polya develops this point of view in more detail in his book, "How to Solve It," first published in 1945; fifth enlarged printing 1948, Princeton University Press.

sorts of problems. The student develops his intelligence by using it, he learns to do problems by doing them. Which kind of problems should a high school student do in order to develop his ability to solve problems?

*A boy or girl of high school age and average ability can solve on a scientific level mathematical problems, but no other kind of problems.* An average boy of fifteen can obviously not acquire the technique or knowledge or judgment needed in treating on a scientific level a problem of biology or history or physics. Yet, if he has a good teacher, he can, after a while, solve a problem of geometric construction or invent by himself the proof of a simple theorem on the level of Euclid, and Euclid's level is fully scientific.

This is the great opportunity of mathematics: mathematics is the *only* high school subject in which the teacher can propose and the students can solve problems on a scientific level. This is so because mathematics is so much simpler than the other sciences. Because of this simplicity, the individual, just as the human race, can arrive so much earlier to a clear view in mathematics than in the other sciences. We should remember that in Euclid's time mathematics was a highly developed science with standards not essentially lower than today, whereas, e.g., biology or medicine was scarcely more than a heap of errors with a few good observations scattered among the rubbish.

3. In my opinion, the first duty of a teacher of mathematics is to use this great opportunity: he should do everything in his power to develop his students' ability to solve problems.

First, he should set his students the right kind of problems: not too difficult and not too easy, natural and interesting, challenging their curiosity, proportionate to their knowledge. He should also allow himself some time for presenting the problem appropriately, so that it appears in the proper light.

Then, the teacher should help his students properly. Not too little, or else there is no progress. Not too much, or else the student has nothing to do. Not ostentatiously, or else the students get disgusted with the problem in the solution of which the teacher had the lion's share. Yet, if the teacher helps his students just enough and unobtrusively, leaving them some independence or at least some illusion of independence, they may experience the tension and enjoy the triumph of discovery. Such experiences may contribute decisively to the mental development of the students.

4. There is a definite technique of helping the students; the author of these lines tried to describe it in his booklet, "How to Solve It."[1] Yet there is a first condition. Nobody can give away what he has not got. No teacher can impart to his students the experience of discovery if he has not got it himself. Therefore, in the opinion of the author, the curriculum for future teachers of mathematics should emphasize, much more than it usually does nowadays, the practical ability to solve not too advanced problems and the methods of solution.[2]

---

1. Princeton University Press. Later books of G. Polya: *Mathematics and Plausible Reasoning* (Princeton University Press), *Mathematical Discovery* (John Wiley & Sons) [cf. vol. 1, pp. 117–18, with section 1 of the present paper], and *Mathematical Methods in Science* (New Mathematical Library).

2. Three courses usually given by the author in Stanford University follow this direction: "Selected Topics from Elementary Geometry," "Mathematical Methods in Science," and "Elementary Mathematics from Higher Point of View."

# 2

# Problem Solving as a Goal, Process, and Basic Skill

## Nicholas A. Branca

THE TERM *problem solving* occurs in many different professions and disciplines and has many different meanings. Troubleshooting, for example, is a form of problem solving; creating new ideas or inventing new products or techniques is another. Although problem solving in mathematics is more specific, it is still open to different interpretations. Activities classified as problem solving in mathematics include solving simple word problems that appear in standard textbooks, solving nonroutine problems or puzzles, applying mathematics to problems of the "real" world, and creating and testing mathematical conjectures that may lead to new fields of study.

*Problem solving*, then, is an all-encompassing term that can mean different things to different people at the same time and different things to the same person at different times. The three most common interpretations of problem solving are (1) as a *goal*, (2) as a *process*, and (3) as a *basic skill.* These three interpretations and some implications they may have for teaching mathematics are explored here.

### Problem Solving as a Goal

Why do we teach mathematics? What are the goals of instruction in mathematics? Educators, mathematicians, and others concerned with these questions frequently cite problem solving as a goal (if not *the* goal) of mathematics learning. Begle (1979) stated, "The real justification for teaching mathematics is that it is a useful subject and, in particular, that it helps in solving many kinds of problems" (p. 143). The same view was expressed by the National Council of Supervisors of Mathematics (1977) in their position paper on basic skills. Similar statements describe mathematics as essentially a problem-solving endeavor (Wirtz 1975) and as a vehicle for generating and exercising problem-solving abilities (Braunfeld 1975). Problem solving has been said to be at the heart of all mathematics (Lester 1977).

When problem solving is considered a goal, it is independent of specific problems, of procedures or methods, and of mathematical content. The important consideration here is that learning how to solve problems is the primary reason for studying mathematics. This point of view influences the tone of the entire mathematics curriculum and has important implications for classroom practices.

## Problem Solving as a Process

Another common meaning of problem solving emerges from its interpretation as a dynamic, ongoing process. In a recent issue of the *Arithmetic Teacher* devoted to problem solving, LeBlanc (1977) stated that "in problem solving an individually acquired set of processes is brought to bear on a situation that confronts the individual" (p. 16). The National Council of Supervisors of Mathematics (1977) defined problem solving as "the process of applying previously acquired knowledge to new and unfamiliar situations" (p. 2).

This interpretation is perhaps best seen in the distinction between the answer students give to a problem and the procedures or steps they use to arrive at the answer. What is considered important in this interpretation is the methods, procedures, strategies, and heuristics that students use in solving problems. These parts of the problem-solving process are its essence and as such become a focus of the mathematics curriculum.

Exactly how one goes about learning the problem-solving process and how one should teach it are not fully understood. Information-processing theory (Newell and Simon 1972) has been used to examine the question of how to teach the problem-solving process (Goldberg 1973; Kantowski 1974; Lucas 1972; Smith 1973), but progress along this line has been slow and difficult (Krulik and Rudnick 1980).

## Problem Solving as a Basic Skill

The last, but by no means the least important, interpretation of problem solving is as a basic skill. What is a basic skill? This question probably has more answers than the question "What is problem solving?" Most of the answers given to the questions about basic skills, however, have included some mention of the concept of problem solving.

During the past few years, various local, state, and national groups, influenced by the movements for back to basics, accountability, and behavioral objectives, have concentrated on defining or evaluating basic skills in mathematics. The reports and position papers of these groups are primarily of two kinds: (1) those concerned with determining minimum skills for evaluation (usually local or state) and (2) those concerned with identifying the basic skills that individuals need to function in our society.

The reports of state agencies concerned with defining and evaluating minimum competencies fall in the first category. Nearly forty states have mandated specific standards for advancement in, or graduation from, their public schools. The remaining states have either legislation pending or legislative or state-board studies under way (Pipho 1978a and 1978b).

There is considerable variability in the mathematical skill areas listed by the different states. Six states list computation or computing as the only skill to be assessed, three list arithmetic, and seventeen simply say "mathematics." Only one state specifically identifies problem solving as a basic skill. The remaining states either leave basic skills unspecified, name a standardized test to be employed, or indicate that the skill areas are a local option.

The most serious shortcomings in the efforts of state and local agencies to implement minimum-competency legislation occur in evaluation. For the most part, basic skills are restricted to skills that can be easily tested by paper-and-pencil tests (preferably using a multiple-choice format). Much of problem solving does not fit this criterion.

The second category of reports and position papers includes those from a number of national, state, and local groups concerned with the mathematics curriculum. These groups include the major curriculum projects, such as the School Mathematics Study Group (SMSG), that were funded during the 1960s and 1970s. As one of its last undertakings, SMSG (1972a) developed a junior high school mathematics curriculum designed to include those—and only those—mathematical concepts that all citizens should know in order to function effectively in our society. The concepts included in this curriculum, *Secondary School Mathematics,* can be considered SMSG's statement of the minimum goals of mathematics education (summarized in SMSG's *Newsletter No. 38* [1972b]), and as such they can be considered a statement of basic skills. Of the twenty-eight chapters produced, it is significant that one, chapter 4, was devoted entirely to problem formulation and a second, chapter 19, was devoted entirely to problem solving.

Another group concerned with curriculum was the National Advisory Committee on Mathematical Education (NACOME), appointed in May 1974 by the Conference Board of the Mathematical Sciences to prepare an overview and analysis of school mathematics education. In their review of some of the newer curriculum projects, NACOME pointed out the extreme difficulty of assessing problem-solving ability. They indicated that answers were not yet available to the fundamental question of whether or not experience with applications improves one's understanding of mathematical ideas or problem-solving ability, and they encouraged teachers at all levels to experiment with available materials to develop balanced mathematical experiences for their students. One of NACOME's major policy recommendations was a warning to teachers, administrators, parents, and the general public against being manipulated into a false choice between structure and problem solving in mathematics programs. Instead, a judicious combination of both elements was advocated for every mathematics program.

In October 1975, the Basic Skills Group of the National Institute of Education (NIE) sponsored and organized a three-day conference in Euclid, Ohio, on basic mathematical skills and learning. Thirty-three mathematicians and educators representing school systems, colleges, universities, and other agencies contributed papers on the theme "What are basic mathematical skills?" The papers ranged from lists of specific mathematics topics arranged by grade levels to treatises on the role of mathematics in society. A number of the papers, including those by Braunfeld, Davis, Gibb, Rising, and Wirtz, emphasized the importance of the skill of problem solving. In summarizing the conference, cochairmen Peter Hilton and Gerald R. Rising echoed the NACOME report when they stated,

> Problems are most efficiently solved by the application of the appropriate theory; and the place where the theory is most likely to be developed is in response to the desire to solve interesting problems. Thus, the two activities of structure building and problem solving are

highly complementary to each other, and indeed, depend on each other in any well-balanced curriculum. [NIE 1975 b, p. 41]

The significance of problem solving as a basic mathematical skill was reinforced by the National Council of Supervisors of Mathematics. During their 1976 annual meeting, more than one hundred members expressed a need for a unified position on basic mathematical skills. A statement of this position would make them more effective leaders in their respective school systems. It would provide both a rationale for a basic mathematics program and direction for implementing it. When they drew up the statement, they listed the skill of problem solving as one of ten basic skills areas.

This position paper (1977) has been reproduced and distributed throughout the states, has filtered from the state level down to the regional and local levels, and is influencing curriculum decision makers.

In interpreting problem solving as a basic skill, one is forced to consider specifics of problem content, problem types, and solution methods. The focus is on the essentials of problem solving that all students must learn, and difficult choices need to be made regarding the problems and techniques to be used.

## Implications for Teaching Mathematics

Although each of the above interpretations of problem solving has its own implications for teaching mathematics, together they give a much broader context from which to draw implications. Two examples illustrate this point. The first comes from the SMSG *Secondary School Mathematics* curriculum mentioned above. The purpose and objectives of the chapter on problem solving (chap. 19) follow:

*Purpose*

All of mathematics is concerned with problem solving. Some problems are theoretical and many are "practical." Problems of various types occur, obviously, throughout secondary school mathematics. In each chapter, we have been concerned with specific techniques developed in the chapter. There are however, certain general strategies and methods that are useful in all types of problems. The major purposes of Chapter 19 are to direct the student's attention to some of these strategies and techniques. The formulation of a mathematical model for a given "real" problem, the analysis of this model, and the interpretation of the analysis in terms of the original problem are stressed. Hopefully, the practice provided by the examples and exercises of this chapter will be reinforced throughout the student's pursuit of his mathematical studies.

*Objectives*

(a) To provide the student with a variety of strategies for problem solving.

(b) To develop some flexibility in the student's approach to problem solving.

(c) To develop techniques for using geometric representations as a way of developing new information about a given situation.

(d) To develop some skill in using a tabular arrangement of given and derived information, to help solve problems.

(e) To develop, for the student, a better understanding of a problem by teaching him to make numerical estimates and testing them in the actual problem. [SMSG 1972 *b*, p. 38]

Each of the three interpretations of problem solving is touched upon in the statement above. Problem solving is interpreted as a goal in the first sentence, as a process in the subsequent focus on strategies and techniques, and as a skill in the list of objectives.

The second example, from the position paper of the National Council of Supervisors of Mathematics (1977), is a statement about problem solving from their list of the ten basic mathematical skills:

> Learning to solve problems is the principal reason for studying mathematics. Problem solving is the process of applying previously acquired knowledge to new and unfamiliar situations. Solving word problems in texts is one form of problem solving, but students also should be faced with nontextbook problems. Problem-solving strategies involve posing questions, analyzing situations, translating results, illustrating results, drawing diagrams, and using trial and error. In solving problems, students need to be able to apply the rules of logic necessary to arrive at valid conclusions. They must be able to determine which facts are relevant. They should be unfearful of arriving at tentative conclusions, and they must be willing to subject these conclusions to scrutiny. [p. 2]

Although the statement concerns problem solving as a basic skill, problem solving is also considered as the principal reason (or *goal*) for studying mathematics and as the *process* of applying knowledge.

We as teachers must realize the importance of problem solving with respect to each of the three interpretations. We must be aware that the students who enter school during this decade will spend the majority of their productive lives solving the problems of the twenty-first century. Although we have no way of knowing what these problems will be like, considering the different interpretations of problem solving can help us prepare for them.

Considering problem solving as a basic skill can help us organize the specifics of our daily teaching of skills, concepts, and problem solving. Considering problem solving as a process can help us examine what we do with the skills and concepts, how they relate to each other, and what role they play in the solution of various problems. Finally, considering problem solving as a goal can influence all that we do in teaching mathematics by showing us another purpose for our teaching.

Each of these interpretations is important, but they are different. When we encounter the term *problem solving,* we should consider which interpretation (or interpretations) is intended. The multiple meanings for the term can easily lead a writer to ambiguity and a reader to misunderstanding. Problem solving has too many facets for us always to look at it from the same angle.

## REFERENCES

Begle, Edward G. *Critical Variables in Mathematics Education.* Washington, D.C.: Mathematical Association of America and the National Council of Teachers of Mathematics, 1979.

Braunfeld, Peter. "Basic Skills and Learning in Mathematics." In *Conference on Basic Mathematical Skills and Learning, Vol. 1,* pp. 23–32. Washington, D.C.: National Institute of Education, 1975.

Davis, Robert B. "Basic Skills and Learning in Mathematics." In *Conference on Basic Mathematical Skills and Learning, Vol. 1,* pp. 42–50. Washington, D.C.: National Institute of Education, 1975.

Gibb, E. Glenadine. "Response to Questions for Discussion." In *Conference on Basic Mathematical Skills and Learning, Vol. 1,* pp. 57–61. Washington, D.C.: National Institute of Education, 1975.

Goldberg, Dorothy J. "The Effects of Training in Heuristic Methods in the Ability to Write Proofs in Number Theory." Unpublished doctoral dissertation, Columbia University, 1973.

Kantowski, Mary Grace. "Processes Involved in Mathematical Problem Solving." Unpublished doctoral dissertation, University of Georgia, 1974.

Krulik, Stephen, and Jesse A. Rudnick. *Problem Solving: A Handbook for Teachers*. Boston: Allyn & Bacon, 1980.

LeBlanc, John. "You Can Teach Problem Solving." *Arithmetic Teacher* 25 (November 1977): 16–20.

Lester, Frank. "Ideas about Problem Solving: A Look at Some Psychological Research." *Arithmetic Teacher* 25 (November 1977): 12–15.

Lucas, John F. "An Exploratory Study on the Diagnostic Teaching of Heuristic Problem-solving Strategies in Calculus." Unpublished doctoral dissertation, University of Wisconsin, 1972.

National Advisory Committee on Mathematical Education. *Overview and Analysis of School Mathematics, Grades K–12.* Washington, D.C.: Conference Board of the Mathematical Sciences, 1975. (Available from the National Council of Teachers of Mathematics, 1906 Association Dr., Reston, VA 22091.)

National Council of Supervisors of Mathematics. "Position Paper on Basic Mathematical Skills." Washington, D.C.: National Institute of Education, 1977. See also *Arithmetic Teacher* 25 (October 1977): 19–22.

National Institute of Education. *Conference on Basic Mathematical Skills and Learning, Vol. 1: Contributed Position Papers.* Washington, D.C.: The Institute, 1975 a.

————. *Conference on Basic Mathematical Skills and Learning, Vol. 2: Working Group Reports.* Washington, D.C.: The Institute, 1975 b.

Newell, A., and H. Simon. *Human Problem Solving.* Englewood Cliffs, N.J.: Prentice-Hall, 1972.

Pipho, Chris. "Minimum Competency Testing in 1978: A Look at State Standards." *Phi Delta Kappan* 59 (May 1978 a): 585–88.

————. "State Activity-Minimal Competency Testing." Working paper of the Department of Research and Information, Education Commission of the States. Denver, Colo., 3 August 1978 b.

Rising, Gerald R. "What Are the Basic Skills of Mathematics?" In *Conference on Basic Mathematical Skills and Learning, Vol. 1.* Washington, D.C.: National Institute of Education, 1975.

School Mathematics Study Group. "Final Report on a New Curriculum Project." *Newsletter No. 36.* Stanford, Calif.: SMSG, 1972 a.

————. "Minimum Goals for Mathematics Education." *Newsletter No. 38*, Section 2. Stanford, Calif.: SMSG, 1972 b.

Smith, J. Phillip. "The Effect of General versus Specific Heuristics in Mathematical Problem-solving Tasks." Unpublished doctoral dissertation, Columbia University, 1973.

Wirtz, Robert W. "Where Do We Go in Mathematics Education?" In *Conference on Basic Mathematical Skills and Learning, Vol. 1.* Washington, D.C.: National Institute of Education, 1975.

*3*

# Heuristics in the Classroom

## Alan H. Schoenfeld

POLYA writes in *How to Solve It* that in studying modern heuristics, one "endeavors to understand the process of solving problems, especially the *mental operations typically useful* in this process" (1945, p. 118). His claims are modest. Polya merely asserts that a better understanding of these general problem-solving strategies "could exert some good influence on . . . the teaching of mathematics." Certainly it can, for teachers' sensitivity to both their own and their students' reasoning processes can be a major variable in classroom dynamics. But many people, myself included, would claim more. Under appropriate circumstances, many students can learn to *use* heuristics, with the result being a demonstrable improvement in their problem-solving performance.

## A Definition and Some Examples

*Heuristic* will be used here to mean a general suggestion or strategy, independent of any particular topic or subject matter, that helps problem solvers approach and understand a problem and efficiently marshal their resources to solve it. Many such strategies exist; the chart in figure 1 provides a large (but still incomplete) sample. For more complete descriptions of these and other heuristics, see Polya (1962; 1965) or Wickelgren (1974).

---

**Some Important Heuristics in Problem Solving**

*Analyzing and Understanding a Problem:*

1. **Draw a diagram** if at all possible.
2. **Examine special cases** to (*a*) exemplify the problem, (*b*) explore the range of possibilities through limiting cases, and (*c*) find inductive patterns by setting integer parameters equal to 1, 2, 3, . . . in sequence.
3. Try to simplify it by using symmetry or "without loss of generality."

---

9

*Designing and Planning a Solution:*

1. Plan solutions hierarchically.
2. Be able to explain, at any point in a solution, what you are doing and why; what you will do with the result of this operation.

*Exploring Solutions to Difficult Problems:*

1. Consider a variety of equivalent problems:
   a) Replace conditions with equivalent ones.
   b) Recombine elements of the problem in different ways.
   c) Introduce auxiliary elements.
   d) Reformulate the problem by (1) a change of perspective or notation, (2) arguing by contradiction or contrapositive, or (3) assuming a solution and determining the properties it must have.

2. Consider **slight modifications** of the original problem:
   a) Choose subgoals and try to attain them.
   b) Relax a condition and then try to reimpose it.
   c) Decompose the problem and work on it case by case.

3. Consider **broad modifications** of the original problem:
   a) Examine analogous problems with less complexity (fewer variables).
   b) Explore the role of just one variable or condition, leaving the rest fixed.
   c) Exploit any problem with similar form, "givens," or conclusions; try to exploit both the result and the method.

*Verifying a Solution:*

1. Use these specific tests: Does it use all the data? Conform to reasonable estimates? Stand up to tests of symmetry, dimension analysis, scaling?
2. Use these general tests: Can it be obtained differently? Substantiated by special cases? Reduced to known results? Can it generate something you know?

Fig. 1

Such strategies are used often and to advantage, no doubt, by experienced problem solvers. To see that using heuristics can be of substantial benefit, let us examine three problems.

**1.** For what values of $a$ does the system of equations
$$\left\{ \begin{array}{l} x^2 - y^2 = 0 \\ (x - a)^2 + y^2 = 1 \end{array} \right\} \quad \text{have 0, 1, 2, 3, 4, or 5 solutions?}$$

**2.** What is the sum of the coefficients of $(x + 1)^{31}$? Prove your answer.

**3.** Given $a$, $b$, $c$, and $d$ real numbers between 0 and 1, prove that
$$(1 - a)(1 - b)(1 - c)(1 - d) > 1 - a - b - c - d.$$

In problem 1, it is very easy to become bogged down in a morass of algebraic computations. However, the suggestion "draw a diagram if at all possible" makes the problem accessible (fig. 2). When we think of $(x - a)^2 + y^2 = 1$ as a circle "sliding" along the $x$-axis, it is easy to see that there are either no solutions, two solutions (at points we can determine as $\pm\sqrt{2}$), three solutions ($|a| = 1$), or four (when $|a| < \sqrt{2}$ and $a \neq 0$, 1, or $-1$).

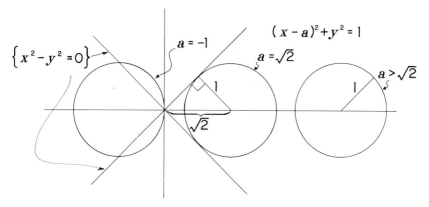

Fig. 2. "Draw a diagram if at all possible."

Problem 2 and similar problems often cause students difficulty (Schoenfeld 1979 a; 1979 b). Yet students who are asked to calculate the sums for the values $n = 1, 2, 3$, or 4 have no difficulty with the problem: the resulting values 2, 4, 8, and 16 lead to the conjecture that the formula is $2^n$, which can be verified by induction. (It can also be obtained by setting $x = 1$, if one knows the binomial theorem.)

Problem 3 can be worked out manually. In fact, the temptation is for students to do so, by multiplying the terms on the left-hand side of the inequality and then showing that $(ab + cb + cd + da - abc - bcd - abd - acd + abcd) > 0$. Students who have solved the problem this way have indicated that they would try the same procedure for a similar problem with $a, b, c, d, e$, and $f$. Yet the problem is easy to solve by first looking at the analogous two-variable problem:

**3'.** Prove $(1 - a)(1 - b) > 1 - a - b$.

Since $(1 - a)(1 - b) = 1 - a - b + ab > 1 - a - b$ (since $a$ and $b$ are positive), we have established (3'). Multiplying both sides of (3') by $(1 - c)$ gives

$$(1 - a)(1 - b)(1 - c) > (1 - a - b)(1 - c)$$
$$= 1 - a - b - c + ac + bc$$
$$> 1 - a - b - c,$$

and multiplying the first and last terms of this inequality by $(1 - d)$ gives

$$(1 - a)(1 - b)(1 - c)(1 - d) > (1 - a - b - c)(1 - d)$$
$$= 1 - a - b - c - d + ad + bd + cd$$
$$> 1 - a - b - c - d,$$

which finishes the argument.

We shall explore some of the many other examples of the utility of heuristic strate-gies in what follows. Using such strategies does improve students' problem-solving performance. However, they must learn (1) how to *select* the appropriate strategies and (2) how to *apply* them, neither of which is as simple as it might seem.

## Some Basic Facts about Heuristics

Unfortunately, figure 1 would most probably be of little value to students if it were given to them as "some suggestions to help in your problem solving." As imposing as the list looks, it is actually very incomplete. Many of the suggestions simply do not contain enough information to be useful to students. To see this, let us examine two of the strategies and some of the problems to which they apply.

### Examine special cases

**4.** Let $P(x) = ax^2 + bx + c$, and $Q(x) = cx^2 + bx + a$. What is the relation-ship between the roots of $P(x)$ and $Q(x)$?

**5.** Prove that in any circle, the central angle that subtends a given arc is twice the inscribed angle subtending the same arc.

**6.** Of all triangles of perimeter 30, which has the largest area?

We saw in problem 2 that "special cases" could mean substituting the values $n = 1, 2, 3, 4$ into the expression and calculating. In problem 4, the quadratic for-mula yields the roots of both $P(x)$ and $Q(x)$ but gives no insight as to the relationship between them. However, if you choose a few easily factored quadratics and calcu-late roots for $P$, it becomes apparent that $Q$'s roots are the reciprocals of $P$'s (table 1).

TABLE 1
"Special Cases" for Problem 4

| $P(x)$ | $Q(x)$ | Roots of $P$ | Roots of $Q$ |
|---|---|---|---|
| $x^2 + 3x + 2$ | $2x^2 + 3x + 1$ | $-2, -1$ | $-1/2, -1$ |
| $x^2 + 3x - 10$ | $-10x^2 + 3x + 1$ | $2, -5$ | $1/2, -1/5$ |
| $6x^2 - x - 2$ | $-2x^2 - x + 6$ | $-1/2, 2/3$ | $-2 \ 3/2$ |

In problem 5 the "special case" that allows for an easy analysis (and a proof of the general result) is when one side of the inscribed angle is a diameter of the circle. In problem 6 examining special cases—short and wide triangles, tall and narrow ones, and some in between—provides the experiential foundation that will make plausible the answer obtained by calculus. Examining another type of special case—holding one side of the triangle fixed and letting the other two sides vary—may lead to the realization that the largest triangle must be isosceles. This is a sub-stantial simplification; it is not hard to take it one step further and show that the larg-est triangle is equilateral.

The point here is that this strategy of **examining special cases** has been used in four very different ways, none of which is clearly implied in its name. "Special cases" is thus a convenient label for those who know how to use it. For those who do not, it is a complex set of behaviors for which careful instruction must be pro-vided. This complexity is by no means restricted to this particular strategy.

### Choose subgoals and try to exploit them

**7.** Find the sum of the whole numbers from 1 to 200 that are multiples of neither 4 nor 9. You may use the fact that the sum of the numbers from 1 to $n$ is $\frac{1}{2}(n)(n + 1)$.

**8.** In how many 0's does the product of the first 100 integers (100!) end?

**9.** Solve the following equation for $x$:

$$\frac{\sqrt{x + 1} + \sqrt{x - 1}}{\sqrt{x + 1} - \sqrt{x - 1}} = 3$$

In problem 7, we establish four separate subgoals: to obtain the sum of the digits from 1 to 200 and the sum of the multiples of 4, 9, and 36, respectively (the last because each multiple of 36 appears as a multiple of both 4 and 9). Then the sum can be found as follows:

$$(1 + 2 + 3 + \cdots + 200) - (4 + 8 + 12 + \cdots + 200)$$
$$- (9 + 18 + 27 + \cdots + 198)$$
$$+ (36 + 72 + \cdots + 180)$$

Pulling out common factors and using the given formula, we get

$$(\tfrac{1}{2})(200)(201) - 4(\tfrac{1}{2})(50)(51) - 9(\tfrac{1}{2})(22)(23) + (36)(\tfrac{1}{2})(5)(6) = 13\ 263.$$

In problem 8, we first note that the number of 0's at the end of a number is the number of factors of 10 it has; since $10 = 5 \cdot 2$, we need only ask how many 5s and 2s appear as factors of 100! In fact, since many more 2s than 5s will appear in the product, the answer (and our subgoal) will be the number of 5s that appear as factors of the first 100 integers, namely twenty-four.

In problem 9, the obvious subgoal is to clear the radicals from the equation. This, in turn, may be done by setting other subgoals: "Can I get all terms involving $\sqrt{x + 1}$ on one side of the equation and all those involving $\sqrt{x - 1}$ on the other?"

As with the previous strategy, we see that the brief description is adequate only as a label; under that label fall a variety of substantially different behaviors, some of which can be quite complex. This is true with many of the strategies listed in figure 1. If we truly expect students to learn them, we must delineate both the different types of behavior subsumed under the label of any strategy and the constituent subskills and offer training in each. This can require substantial instruction. In addition, that instruction should be more than merely *descriptive;* that is, it must consist of more than mere presentations of problems that are successfully solved by the strategies. It must be *prescriptive,* showing the students how and when to employ each of the strategies over a wide variety of problems.

For example, consider the solution to problem 3 above, which was reached by "exploiting a similar problem with fewer variables." How do we know to look at the two-variable problem? Should we try to solve it first or see if we'll be able to use it if we do solve it? And which will we try to exploit: the method of solution or the result of the two-variable case? Unless these considerations are explained to students and we give them a reasonable procedure for making such decisions themselves, it is unlikely that they will succeed in using the strategy.

## In Addition to Heuristics, What Else?

Even if we could provide students with training in each of the individual strategies listed in figure 1, that alone might result in little or no difference in their overall problem-solving performance. There are so many strategies one *might* use on a problem that, without a reasonable way to select an appropriate approach to a problem, students may run out of time (or patience) before they select the "right" strategy for it.

Perhaps the following analogy will be helpful. We can think of the large collection of strategies in figure 1 as a set of keys, a small number of which may serve to unlock any particular problem. If problem solvers have only the time to try a few of those keys, they may fail to unlock the problem—even if the right key was at their disposal all along. Selecting the key is as important as knowing how to use it.

There are two ways to help students select the appropriate strategies for solving individual problems. The first is to point out "cues" in the very form of problems themselves, which may suggest that a particular approach is appropriate. For example, consider these two problems:

**10.** What is the sum of the series $1 + 3 + 5 + 7 + \cdots + (2n - 1)$?

**11.** How many straight lines can be drawn through $n$ points?

In both of these problems, as in problem 2, the integer parameter, $n$, is a cue telling the problem solver that an inductive argument might be appropriate. It does not guarantee it, of course, nor does it mean that induction will be the best way to solve a problem (there is a simple combinatorial argument for problem 11). But it does suggest that if no other options are more immediately attractive, problem solvers might wish to calculate some values and look for a pattern or to see what happens as they pass from the value of $n$ to $n + 1$. Similarly, there is a clear cue in the next problem:

**12.** Prove that for $x$, $y$, and $z$ positive,
$$(x^2 + 1)(y^2 + 1)(z^2 + 1) \geq 8xyz.$$

The symmetric role of the three variables, parallel to the roles of $a$, $b$, $c$, and $d$ in problem 3, should lead the problem solver at least to *consider* the strategy of fewer variables. Here the idea is to prove that $x^2 + 1 \geq 2x$ and to take the product of the inequality with itself three times (once each with $x$, $y$, and $z$).

Another such cue is the word *unique;* asked to prove that anything is unique (even if I don't know what it is), I will assume that there are two such things and try to derive a contradiction. Contradiction can also be cued by either explicit or implicit negatives. These problems are examples of this:

**13.** Show there is no real number $x$ such that
$$x^{12} + x^8 + x^2 + 6 = 0.$$

**14.** Prove there is an infinite number of primes.

Problem 13 begs for the question, What if I had a real $x$; what could I say about $x^{12} + x^8 + x^2 + 6$? In problem 14, *infinite* says "not finite"; the question to ask is, What if there were only finitely many primes?

Many such cues occur in mathematics (for example, the word *largest* suggests maximization using the derivative in calculus). The point is that the teacher must be aware of them and be willing to share them. Whenever you solve a problem, even a routine one, you should not say, "This is how it's done," but rather ask yourself, "Why did I do it that way?" and share that insight with your students.

For the vast majority of problems, we do not know—or have not yet recognized—the appropriate cues. What can we offer our students then? The first approximation was offered by Polya in his "rules of preference" (1965, vol. 2, p. 93). But this can be taken further: general strategies can be listed, *all other factors being equal,* in groupings that are most likely to be appropriate at particular stages of the problem-solving process and (within those stages) compatible with the rules of preference. This, in fact, was the organization underlying the chart in figure 1 and is what makes the overwhelming list of problem-solving strategies contained there somewhat manageable. When reading the problem and trying to get a feel for it, for example, the problem solver is most likely to need the strategies listed under "Analyzing and Understanding a Problem." If the problem proves to be difficult, a variety of explorations may be appropriate. The strategies under "Exploring Solutions to Difficult Problems" are listed, in general, in the order of their usefulness. In terms of the analogy that opened this section, this kind of grouping serves to present the student, at any stage in problem solving, with the subset of keys most likely at that point (with all other factors being equal) to unlock the problem.

The overall problem-solving process is, of course, far more complex than can be summarized in a list of strategies like the one given in figure 1—and the fate of such useful lists, unfortunately, is that students tend to lose them! The list is meant to serve as a reference and as a framework for solving problems by means of heuristic strategies. The process of problem solving, including illustrations of the kinds of tactical decisions (regarding the allocation of problem-solving resources) that problem solvers make all the time, can and should be discussed often in the classroom. We shall examine a sample classroom discussion in the latter part of the next section.

## In the Classroom

The *problem-solving process* is one of the most important aspects of mathematics with which teachers should be concerned. Very few adults ever have an overt use for, say, the quadratic formula or for proving a theorem in geometry; what they can and should have, as a consequence of their education, is the ability to reason carefully and to make intelligent and efficient use of the resources at their disposal when confronted with problems in their own lives.

There are many other reasons for focusing on the problem-solving process in the classroom. Certainly a class in which the students are helping the teacher work out problems and (at least apparently) contributing actively to their solutions is more likely to be dynamic and motivating than one that follows the classical "show and drill" mode. Explaining to students where arguments come from—or better yet, working the arguments out with them when possible—can help to demystify mathematics and allow the students to approach it with less fear and trepidation.

Now let's deal with teaching heuristics in the context of a problem-solving course. Such a course offers a unique opportunity to pull things together for students, focusing on problem solving rather than on subject matter. But we should not think of problem solving as a separate skill to be taught in a separate course. Every hour in the classroom presents opportunities for showing students how to think mathematically. Most of the suggestions in what follows can, and should, be used in everyday instruction.

### The class format

Since the emphasis of instruction in problem solving is on the *process* of solving problems, the format of the class must reflect and encourage this. Clearly a certain amount of time must be set apart for straightforward presentations: outlining problem-solving strategies, establishing the appropriate context for classwork, providing background materials, offering concise summaries, and so on. But most of the class time should be spent in solving problems. This can be profitably done in two ways:

1. *The discussion format.* Here the teacher serves as conductor to the student orchestra of suggestions, gently guiding the students through the problem-solving process by using their suggestions as often as possible and training them to use the strategies. (An outline of a sample discussion will follow later.)

2. *The small-group approach.* The class can be divided into groups of four or five students each. These groups work together on an assignment of two or three problems for fifteen or twenty minutes, and during this time the teacher circulates through the classroom, giving help when absolutely necessary. When the groups have either solved the problems or made as much progress as is likely, the class returns to the discussion format.

It should be noted that the amount of material covered in any class session is usually quite small; it may be that only four or five problems can be discussed in a one-hour class. The teacher should not find this disturbing; it is a natural consequence of paying attention to the process of problem solving. Our goals are quite different from having students merely see and appreciate the mathematics involved. (After all, it can take many weeks to learn to make even a mediocre copy of a Rembrandt pencil sketch, although one can study and appreciate it in a short time!)

### Motivation

If we really expect students to take problem-solving strategies seriously, we must convince them that something is to be gained from studying them. Perhaps the easiest way to do this is to be armed at the beginning of a course (or any particular class session) with some problems that dramatically demonstrate the impact of heuristics—such as the first three problems given in this essay. The student who sees a diagram unlock problem 1 after he or she has struggled with an algebraic solution is more likely to use diagrams in the future. Another such problem (if you are willing to take the risk) is the following:

   **15.** John's wealthy and eccentric uncle left him in his will a triangular plot of land whose sides measured 25, 50, and 75 yards. If the land is valued at $25 a square yard, what is the value of John's inheritance?

The answer can be obtained through Heron's formula. However, the student who draws the plot of land is likely to see, without doing any calculations, that the ''triangle'' collapses. There is, actually, a serious point to be learned: one should not get immersed in complex calculations until they are clearly appropriate.

Similarly, problem 2 serves to make a dramatic point. Students who have had difficulty with it and then see how easy it can be with the help of the inductive strategy may be tempted in the future to substitute values for an integer parameter in other problems that have them. There are other such dramatic examples, many of which can be found in Wickelgren (1974). Just one word of warning: we should not let students leave the classroom thinking that the impact of heuristics will always be as dramatic as it is in these problems.

### Teaching any particular strategy

If we really expect students to use a heuristic strategy, we must teach it with the same degree of seriousness we would devote to any other mathematical technique, be it a procedure for solving mixture problems or the use of the derivative in maximization problems. This means, for example, that we should have a collection of problems ready to exemplify the use of the strategy. Given that these problems are discussed at length, a collection of, say, eight to a dozen is all that is necessary. Problems 2, 4, 5, and 6 provide a solid basis, for example, for a discussion of the strategy "examine special cases." Other sources of problems are Bryant et al. (1965); Burkill and Cundy (1961); Dynkin et al. (1969); Eves and Starke (1957); Lidsky et al. (1963); Polya and Kilpatrick (1974); Rapaport (1963); Shklarsky et al. (1962); Trigg (1967); and Yaglom and Yaglom (1967).

For motivation, one or two problems solvable by a particular strategy might be assigned for the class session devoted to that strategy. The class, using the discussion format, should work through the problems, with the teacher (at the end) highlighting the use of the strategy in the solutions and providing a description of the strategy in general. For more practice, the class is given three or four other problems and divided into small groups to work on them. After these problems have been discussed, the teacher pulls things together with a summary. Then a number of other problems are assigned for the next session—but not all requiring the same strategy. Some of the problems should be solvable by methods studied earlier in the course and perhaps one or two by a method soon to be studied. And for each problem that is discussed, the teacher and students should focus not only on how the approaches worked but on why they were appropriate for use on the problem.

### A sample discussion

To illustrate what is meant by focusing on the *process* of solution, here is the outline of a discussion that might take place about one particularly malleable problem. The problem itself is a variant of one discussed by Polya in *How to Solve It* (1945, pp. 23–25), modified to allow for a greater variety of possible approaches.

**16.** You are given the triangle *T* in figure 3, with its longest side as its base. Prove that a square, *S*, can be inscribed in *T*; that is, that there is a square whose four vertices lie on the sides of *T*.

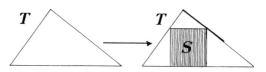

Fig. 3. "Try first to solve an easier, related problem."

The strategy to be highlighted in class is this:

*If you cannot solve the proposed problem, try first to solve an easier, related problem. Then try to exploit your solution.*

The discussion can be started by asking the class if the desired result seems plausible. After some examination (which may lead to the most obvious solution), it does indeed appear to be reasonable. Writing the foregoing heuristic on the blackboard, we can ask: "What is an easier, related problem you might consider?" The answers might be like these:

1. Prove we can inscribe a rectangle in *T*.
2. Prove we can inscribe a circle in *T*.
3. Prove we can inscribe an equilateral triangle in *T*.
4. Prove that the triangle *T* can be circumscribed around a square.

After the class has generated these responses, the teacher and class should examine them. Two critical questions should be considered: (*a*) Can we solve the easier related problem? and (*b*) Will solving it help us solve the original problem? Consider the following points with regard to the four possibilities generated above: (1) looks easy to solve, and may help us; (2) is a known theorem, but there is no assurance it will be helpful; (3) is problematic, and we are not sure it would be useful; (4) is (apparently) a mere restatement of problem 16 and probably will not be considered useful.

If the students see how to exploit the fourth possibility for an easier related problem, they might be given the choice of pursuing either this or the first, the two most likely approaches. Otherwise the consensus will probably be to pursue the first. *Note that the selection of an appropriate approach is critical to solving this problem. Its role should not be underplayed; the ability to do this is what will make or break problem solvers!*

So we ask if anyone can inscribe a rectangle in *T*. If we are lucky, someone will say yes and will construct it by drawing a line parallel to the base and dropping perpendiculars to the base from where the line intersects the sides (fig. 4). If not, we can lead the class to it.

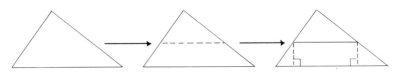

Fig. 4. Constructing a rectangle in a triangle

We should encourage the class to finish the argument from there. (See Wickelgren's technique for guiding people through solutions in *How to Solve Problems* [1975].) We might ask, for example, "Since rectangles are easier than squares, can we inscribe another rectangle in *T*?" Yes. "Any more?" Yes. (See fig. 5.)

And now the important question: How can we describe the variety of rectangles that are possible? Easing the condition of the problem gives us many rectangles: short wide ones, tall narrow ones, and so on. And what happens as we move from the short wide rectangles at the bottom of figure 5 to the tall narrow ones at the top?

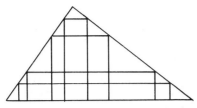

Fig. 5. The variety of rectangles in a triangle

Aha! At the bottom, the rectangles are longer horizontally than vertically, and at the top they are longer vertically. At a point somewhere between, therefore, base equals height—and there's the square we were looking for.

We have now solved the problem, at least as far as the square's existence is concerned, but we are far from finished with it. After all, the class has generated a number of approaches to the problem; we should see if any might be useful. Their second and third suggestions had seemed implausible—but might the class convert the fourth to something more useful? If no one answers, then the suggestion "What if we could put a triangle similar to $T$ around a square?" might stimulate a solution such as this:

Drop the perpendicular to the base of $T$. If its length is $h$, construct a square of side $h$. Now copy the angles $(90 - \alpha)$ and $(90 - \beta)$ at the left and right upper vertices of the square, and extend these sides and the base of the square until they intersect (fig. 6).

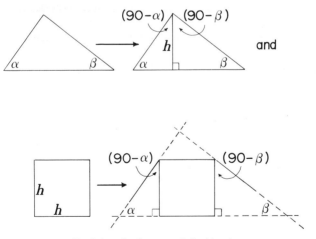

Fig. 6. Locating the square in the triangle

We note that the base angles of the constructed triangle are $\alpha$ and $\beta$, and so it is similar to the original. Since it contains an inscribed square, the original must contain one also.

We have now used the heuristics to solve the problem, but there is still more to be done: the students should see how the strategy works in general. An important question should be considered: How do we find easier related problems? Generally, we can find them by assuming more than what was given in the problem or by trying to prove somewhat less than the original problem asks for. The problem we were given has *two* conditions: we wanted (1) a square, and (2) all four vertices of the square to lie on the sides of *T*. The class had *weakened* the first condition by asking for a rectangle instead of a square; they might instead have weakened the second. Can we find a square with only three vertices on *T*? Of course (fig. 7). And varying this square leads to yet another solution.

Fig. 7. Finding a square with only three vertices on *T*

Finally, we might show the class that an entirely different strategy leads to a solution of a completely different nature. What if we did have a square that satisfied the conditions of the problem? From figure 8, we see that if *B* is the base of the triangle *T* and *s* is the side of the inscribed square, then

$$s = \frac{B}{1 + \cot\alpha + \cot\beta}$$

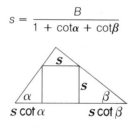

Fig. 8. An analytic solution

Once the denominator is shown to be nonzero, we have an analytic solution of the problem.

Have we been merely flogging a dead horse in this extensive discussion of one problem? Absolutely not; the ability to generate easier and useful problems related to a given problem is valuable. The knowledge that different options exist in problem solving and that one can choose from among them more or less rationally is important; watching the mechanics of thinking through such problems is beneficial. If anything, answering problem 16 by presenting a neatly packaged version of any of our arguments to a class is to abuse the problem. For all but the few students who can do all of the above on their own, an awareness of these processes is the first step in learning to use them.

## Conclusion

As three acknowledged experts in problem-solving theory describe them (Newell, Shaw, and Simon 1960, p. 257), "Heuristics [are] things that aid discovery. Heuristics seldom provide infallible guidance. . . . Often they 'work,' but the results are variable and success is seldom guaranteed.'' Problem-solving strategies are both complex and subtle, as we saw in the second and third sections of this essay. These are good reasons to be cautious and careful but no cause for pessimism. In fact, recent developments in mathematics education, psychology, and artificial intelligence give us cause for tremendous optimism. We are, on many fronts, coming to a more complete understanding of the problem-solving processes of both ''experts'' and students and discovering how to have students use expert strategies successfully. The short-term benefits of paying attention to the problem-solving process in the classroom are at least a demystification of mathematics and a more lively classroom. We shall be counting the long-term benefits happily for many years.

## REFERENCES

Bryant, Steven J., George E. Graham, and Kenneth G. Wiley. *Non-Routine Problems in Algebra, Geometry, and Trigonometry*. New York: McGraw-Hill Book Co., 1965.

Burkill, J. C., and H. M. Cundy. *Mathematical Scholarship Problems*. Cambridge: Cambridge University Press, 1961.

Dynkin, E. B., S. A. Molchanov, A. L. Rozental, and A. K. Tolpygo. *Mathematical Problems: An Anthology*. Translated by Richard A. Silverman. New York: Gordon & Breach, 1969.

Eves, Howard, and Emory P. Starke. *The Otto Dunkel Memorial Problem Book*. Washington, D.C.: Mathematical Association of America, 1957.

Goldberg, Dorothy. ''The Effects of Training in Heuristic Methods on the Ability to Write Proofs in Number Theory.'' (Doctoral dissertation, Columbia University, 1974.) *Dissertation Abstracts International* 34 (1974): 4989B. (University Microfilms no. 75-7836)

Hajos, G., G. Neukomm, J. Suranyi, and J. Kurschak. *The Hungarian Problem Book*. Translated by Elvira Rapaport. Washington, D.C.: Mathematical Association of America, 1963.

Kantowski, Mary G. ''Processes Involved in Mathematical Problem Solving.'' *Journal for Research in Mathematics Education* 8 (May 1977): 163–80.

Landa, Lev. ''The Ability to Think: Can It Be Taught?'' *Soviet Education*, vol. 18, no. 5, March 1976.

Larkin, Jill H. ''Teaching Problem Solving in Physics: The Psychological Laboratory and the Practical Classroom.'' In *Proceedings of the Carnegie-Mellon Conference on Problem Solving and Education*, edited by F. Reif and D. Tuma. Hillsdale, N.J.: Lawrence Erlbaum Associates, 1979.

Lidsky, V., L. Ovsyannikov, A. Tulaikov, and M. Shabunin. *Problems in Elementary Mathematics*. Translated by V. Volosov. Moscow: Mir Publishers, 1963.

Lipson, Stanley H. ''The Effects of Teaching Heuristics to Student Teachers in Mathematics.'' (Doctoral dissertation, Columbia University, 1972.) *Dissertation Abstracts International* 32 (1972): 2221A. (University Microfilms no. 72-30,334)

Lucas, John. ''An Exploratory Study on the Diagnostic Teaching of Heuristic Problem Solving Strategies in Calculus.'' (Doctoral dissertation, University of Wisconsin, 1972.) *Dissertation Abstracts International* 32 (1972): 6825A. (University Microfilms no. 72-15,368)

Newell, Allen, and Herbert Simon. *Human Problem Solving*. Englewood Cliffs, N.J.: Prentice-Hall, 1972.

Newell, Allen, John C. Shaw, and Herbert Simon. ''A Report on a General Problem Solving Program.'' In *Proceedings of the International Conference on Information Processing*. Paris: UNESCO, 1960.

Polya, George. *How to Solve It*. Princeton, N.J.: Princeton University Press, 1945.

———. *Mathematical Discovery*. 2 vols. New York: John Wiley & Sons, 1962, 1965.

Polya, George, and Jeremy Kilpatrick. *The Stanford Mathematics Problem Book*. New York: Teachers College Press, 1974.

Rubenstein, Moshe. *Patterns of Problem Solving*. Englewood Cliffs, N.J.: Prentice-Hall, 1975.

Schoenfeld, Alan H. ''Presenting a Strategy for Indefinite Integration.'' *American Mathematical Monthly* 85 (October 1978): 673–78.

———. ''Can Heuristics Be Taught?'' In *Cognitive Process Instruction*, edited by John Lochhead, pp. 315–38. Philadelphia: Franklin Institute Press, 1979 a.

———. ''Heuristic Variables in Instruction.'' In *Task Variables in Mathematical Problem Solving*, edited by Gerald A. Goldin and C. Edwin McClintock, chap. 10B. Columbus, Ohio: ERIC, 1979 b.

Shklarsky, D. O., N. M. Chentzov, and I. M. Yaglom. *The U.S.S.R. Olympiad Problem Book*. Edited by Irving Sussman and translated by John Maykovich. San Francisco: W. H. Freeman & Co., 1962.

Smith, John P. "The Effects of General versus Specific Heuristics in Mathematical Problem Solving Tasks." (Doctoral dissertation, Columbia University, 1973.) *Dissertation Abstracts International* 33 (1973): 2400A. (University Microfilms no. 73-26,637)

Trigg, Charles W. *Mathematical Quickies*. New York: McGraw-Hill Book Co., 1967.

Wickelgren, Wayne. *How to Solve Problems*. San Francisco: W. H. Freeman & Co., 1974.

Wilson, James. "Generality of Heuristics as an Instructional Variable." (Doctoral dissertation, Stanford University, 1967.) *Dissertation Abstracts* (1967): 2575A. (University Microfilms no. 67-17,526)

Yaglom, E., and H. Yaglom. *Challenging Mathematical Problems with Elementary Solutions*. San Francisco: Holden-Day, 1967.

## FOR FURTHER READING

Brown, John S., and Richard R. Burton. *Diagnostic Models for Procedural Bugs in Basic Mathematical Skills*. Report 3669. Cambridge, Mass.: Bolt, Beranek & Newman, 1977.

Campbell, D. T., and J. C. Stanley. *Experimental and Quasi-Experimental Designs for Research*. Chicago: Rand McNally & Co., 1963.

Kilpatrick, Jeremy. "Variables and Methodologies in Research in Problem Solving." Paper read at the research workshop on Problem Solving in Mathematics Education, May 1975, at the University of Georgia. Mimeographed.

Klahr, David, ed. *Cognition and Instruction*. New York: Halsted Press, 1976.

Kleinmuntz, Benjamin. *Problem Solving: Research, Method, and Theory*. New York: John Wiley & Sons, 1966.

Reif, Fred, and David Tuma, eds. *Proceedings of the Carnegie-Mellon Conference on Problem Solving and Education*. Hillsdale, N.J.: Lawrence Erlbaum Associates, 1979.

Webb, Norman L. "An Exploration of Mathematical Problem Solving Processes." Unpublished doctoral dissertation, Stanford University, 1975.

# 4

# Posing Problems Properly

## Thomas Butts

$F$OR ME, and I suspect the same is true for many other people, the real joy in studying mathematics is the feeling of exhilaration that comes from solving a problem—the tougher the problem, the greater the satisfaction. But what factors initially motivate someone to *want* to solve a problem? Answers to this question can range from personal curiosity to the fear of the consequences if the solution is not handed in tomorrow, but a prime consideration has to be the manner in which the problem is posed. Examine the following problems:

*Problem 1.* Let $d(n)$ denote the number of positive divisors of the integer $n$. Prove that $d(n)$ is odd if and only if $n$ is a square.

*Problem 2.* Which positive integers have an odd number of factors? (Justify your answer.)

*Problem 3.* Imagine $n$ lockers, all closed, and $n$ men. Suppose the first man goes along and opens every locker. Then the second man goes along and closes every other locker beginning with #2. The third man then goes along and changes the state of every third locker beginning with #3 (i.e., if it's open, he closes it, and vice versa). If this procedure is continued until all $n$ men have passed by all the lockers, which lockers are then open?

These three problems are, in fact, different formulations of the same problem. (To see the equivalence of problems 2 and 3, examine a particular locker, number 12, for example. Locker 12 is touched by men 1, 2, 3, 4, 6, 12—i.e., the factors of 12. Since a locker is alternately opened and closed, lockers whose numbers possess an odd number of factors will be open in the end.)

The first version is posed in typically dry, mathematical style. The second, less ponderous, is given as a question to answer rather than as a statement to prove. The third asks this mathematical question in a very picturesque manner. I would argue that the phrasing of the third version (and, to a lesser extent, the second) would significantly motivate the potential solver to tackle the problem.

This paper, then, will be concerned with the question of how to pose (or "repose") a problem to maximize this source of motivation. Before carefully investigating the art of problem posing, let us first briefly discuss the different types of mathematical problems.

# Types of Problems

We shall partition the set of mathematical problems into five arbitrarily titled subsets:

1. Recognition exercises
2. Algorithmic exercises
3. Application problems
4. Open-search problems
5. Problem situations

A brief summary of each category follows.

### Recognition exercises

This type of exercise typically asks the solver to recognize or recall a specific fact, definition, or statement of a theorem.

*Example 1:* Which of the following are polynomials?
(a) $x^3 + 3x + 2$, (b) $x^3 + 3\sqrt{x} + 2$, (c) $x^3 + \sqrt{3}x + 2$, (d) $x^3 + 3/x + 2$, (e) 2

*Example 2:* The line segment joining a vertex of a triangle to the midpoint of the opposite side is called a __?__ .

*Example 3:* If $a$, $b$, $c$ are real numbers and $a > b$, then $ac > bc$. True or false?

### Algorithmic exercises

As the adjective *algorithmic* implies, these are exercises that can be solved with a step-by-step procedure, often a numerical algorithm.

*Example 4:* Compute $16 + 4 \cdot (-2) - (6 \div 3)$.
*Example 5:* Solve $2x^2 - 3x - 5 = 0$.
*Example 6:* Find the center and the radius of the circle
$$x^2 - 2x + y^2 + 4y = 4.$$

### Application problems

Application problems involve applying algorithms. Traditional word problems fall in this category in that their solution requires (a) formulating the problem symbolically and then (b) manipulating the symbols according to various algorithms.

*Example 7:* If the length and width of a rectangle are each increased by 20%, by what percent is the area increased?

*Example 8:* A bag of nickels, dimes, and quarters contains 435 coins worth $43.45. There are three times as many dimes as quarters. How many of each type of coin are in the bag?

A high percentage of all exercises and problems in elementary, secondary, and beginning college textbooks fall in these first three categories. The distinguishing feature of these problems is that *a strategy for solving the problem is contained within its statement*. The hurdle to overcome, then, is to translate the written word into an appropriate mathematical form so that suitable algorithms can be applied.

## Open-search problems

An open-search problem is one that does not contain a strategy for solving the problem in its statement. The problem given at the beginning of this essay falls in this category. Typically such problems are phrased, "Prove that _____," "Find all _____," or "For which _____ is _____," but many other, more interesting variations are possible.

Two other open-search problems (favorites of mine) are these:

*Example 9:* Let $S_n = \{1, 2, 3, \ldots n\}$. For what type of integers, $n$, can $S_n$ be partitioned into two subsets so that the sum of the elements in each subset is the same? For example, $S_7 = \{1, 2, 3, 4, 5, 6, 7\}$ can be partitioned into $\{1, 6, 7\}$ and $\{2, 3, 4, 5\}$.

*Example 10:* How many different triangles with integer sides can be drawn having a longest side (or sides) of length 5 cm? 6 cm? $n$ cm? In each case, how many of the triangles are isosceles?

Though a higher percentage of problems in upper-level college texts are in this category, they are often unimaginatively phrased (e.g., "Prove theorem 2.3"). One of the great misconceptions involving the open-search problem is that it must be concerned with sophisticated mathematical concepts. A tragedy of some instructional methods at the elementary and secondary level is that these students are not given the opportunity to solve open-search problems that depend on the most elementary concepts. The experience of finding a pattern, so fundamental in the study of mathematics, should be offered at all levels of mathematical training. One solution to the problem given in example 9 depends only on the ability to add positive integers and to perceive that the sum of an odd number of odd integers equals an odd integer.

## Problem situations

This category is best typified by Henry Pollak's exhortation: "Instead of telling students, 'Here is a problem, solve it,' tell them, 'Here is a situation, think about it.' " Included in this subset, then, are not problems per se but situations in which one of the crucial steps is to identify the problem(s) inherent in the situation whose solution will improve it. This approach, espoused by the Unified Science and Mathematics for Elementary Schools program (USMES) and others, is part of comprehensive problem solving.

*Example 11:* Design a parking lot. Possible problems to consider could include the following. There are many, many others.

( *a* ) How large should each space be?

( *b* ) At what angle should each space be placed?

( *c* ) How much should be charged per car per hour if one wishes to make a 10% profit?

Clearly a gray area exists for each category, but these subsets can serve as a basis for our discussion on the art of posing problems.

## Some Suggestions for Posing Problems

In this section we shall offer some specific suggestions on posing exercises and problems of the first four types: recognition, algorithmic, application, and open-search. The key step in the fifth type, problem situations, is to identify the problems, which usually fall into the application and open-search categories. Some of these suggestions will, of course, apply to several types of problems; they are placed in the category of greatest impact.

### Recognition exercises

Since the main function of recognition exercises is to test the recall of definitions, basic facts and theorems, and so on, these exercises are usually posed in the *true-false*, *multiple-choice*, *fill-in-the-blank*, or *matching* format. Teachers sometimes hesitate to give such exercises for fear students will memorize without understanding. The "give an example of" problem is often effective in such situations because of its nonspecific answers. With possible prior warning, one may also give such a problem with no solution.

Example 12: Give, if possible, an example of each of the following:

(a) A proper fraction greater than $3/4$

(b) A polynomial of degree 5 with four terms

(c) A triangle, one of whose altitudes coincides with one of its medians

(d) The equation of a circle that is tangent to both coordinate axes

(e) A differentiable function $f(x)$ that is not continuous at $x = 2$

(f) Two three-dimensional vectors, $\vec{A}$, $\vec{B}$, for which $\vec{A} \circ \vec{B} = |\vec{A} \times \vec{B}|$

Such exercises are sometimes more time-consuming to grade as test items, but as problems posed during class they can generate a variety of responses and often stimulate worthwhile discussion.

### Algorithmic exercises

The current interest in basic skills has placed renewed emphasis on algorithmic and application problems. Computational proficiency, in its broadest sense, requires drill and practice; the challenge is to make it interesting. Two suggestions are offered in this vein.

1. *Give a sequence of algorithmic exercises with a purpose.*

Example 13: Choose two whole numbers. Find their sum and nonnegative difference. Add these results. Any observations?

Example 14: Compute:

(a) $\dfrac{1}{1 \cdot 2} + \dfrac{1}{2 \cdot 3}$,

(b) $\dfrac{1}{1 \cdot 2} + \dfrac{1}{2 \cdot 3} + \dfrac{1}{3 \cdot 4} + \dfrac{1}{4 \cdot 5}$,

(c) $\dfrac{1}{1 \cdot 2} + \dfrac{1}{2 \cdot 3} + \ldots + \dfrac{1}{99 \cdot 100}$ . Generalize.

*Example 15:* Multiply:

(a) $(1 + x)(1 - x + x^2)$,
(b) $(a + 2b)(a^2 - 2ab + 4b^2)$,
(c) $(3y - 2w^2)(9y^2 + 6yw^2 + 4w^4)$

*Example 16:* Find the areas of the shapes in figure 1.

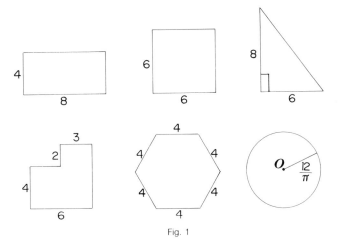

Fig. 1

Do these figures have anything in common? Do you have any observations?
Completing magic squares and magic triangles is another example illustrating this suggestion.

*2. Give the reversal of a familiar problem*

Ask the problem in the "opposite way." Some classic examples of standard problems whose reversals are also standard include the following:

(a) Multiply polynomials; factor polynomials.

(b) Given a polynomial, find its roots; given a set of roots, find a polynomial with those roots.

(c) Differentiation; integration.

Note that the reversal of a problem often has more than one solution. Some less familiar examples of reversal problems are these:

*Example 17:* Find three arithmetical problems whose solution is 13. (You can put as many boundary conditions as you wish on such a problem—e.g., specifying that at least five numbers and at least one +, x, and ÷ sign must be used.)

*Example 18:* Find three solids whose surface area is 60.

*Example 19:* Find a cubic polynomial that has a relative minimum at $(-1, 2)$, a relative maximum at $(2, 7)$, and an inflection point at $(1, -2)$.

*Example 20:* Write 525 as the sum of consecutive integers in as many ways as possible.

*Example 21 (Cryptarithms):* Although not strictly an algorithmic exercise, a cryptarithm is an excellent vehicle for testing the understanding of numerical algorithms. Three of my favorites follow:

$D = 5$. *Other letters stand for a unique digit.*

| (*a*) | DONALD | (*b*) | HOCUS | (*c*) | ABC |
|---|---|---|---|---|---|
| | +GERALD | | +POCUS | | ×BAC |
| | ROBERT | | PRESTO | | ***C |
| | | | | | **A |
| | | | | | ***B |
| | | | | | ****** |

### Application problems

Since the universal complaint against so-called word problems is their artificiality, the most obvious suggestion for improving application problems is to make them realistic. The joint MAA-NCTM *Sourcebook on Applications* gives several criteria for a "good example" in application problems, three of which are worthy of mention here:

1. Data should be realistic, both in the type of information known and the numerical values used. (A problem that asks for the length of a room given its perimeter or area would be artificial.)
2. The "unknown" in the problem should reasonably be expected to be unknown in reality. (The canonical age problem fails this test miserably.)
3. The answer to the problem should be a quantity someone might plausibly have a reason to seek.

Four examples, taken from the *Sourcebook*, illustrate these maxims:

*Example 22:* A six-lane track for running footraces is in the shape of a rectangle whose length is 1.5 times its width with a semicircle on each end. Each lane is to be 1 meter wide. What is the length and width of the rectangle if the inside track is to be 1500 meters long? For a 1500-meter race, the inside runner would start at the finish line. Where should the runners in the other five lanes start?

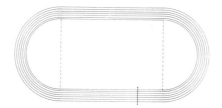

*Example 23:* If an 8-inch pizza serves two, how many should two 12-inch pizzas serve?

*Example 24:* A teacher wishes to rescale the scores on a set of test papers to "improve" everyone's grade. The maximum score is still 100, but 56 becomes a 70 on the new scale. Your score was 75. What does it become on the new scale?

*Example 25:* Suppose you wish to buy a new car and you feel you can afford, at most, $100 a month. If used-car loans are made for a maximum of 30 months at 14% interest, what price car should you consider?

This seems the appropriate place to mention the posing of problems containing insufficient or extraneous data. The ability to discern the data necessary to solve a problem is certainly crucial in real-world problem solving. Consequently, posing problems demanding a critical analysis of the data is to be commended.

*Example 26:* Two sides of a triangle have lengths 4 cm and 6 cm. Find the perimeter and area of the triangle.

Although the data are clearly insufficient to determine a unique solution, one should seek a solution of the form "any real numbers, $P$, $A$, satisfying $12 < P < 20$, $0 < A \leq 12$" rather than accepting "no solution." Problems possessing more than one solution are perfectly reasonable problems.

One could give a hokey problem with extraneous data (a person's age, shoe size, birth date, etc.), but the best application problems require the solvers to gather their own data—a type of problem more properly categorized as a problem situation.

### Open-search problems

Since an open-search problem does not contain a strategy within its statement, solving such a problem requires higher-level thinking. The most important function of open-search problems is to *encourage guessing* (Polya says, "Let us teach guessing"); writing proofs will follow. Especially with an emphasis on guessing, such problems can, and should, be used at all mathematical levels.

The fundamental axiom of posing an open-search problem is—

*Pose the problem in a manner that requires the solver to guess the solution.*

Though this axiom cannot always be followed, all too often a good problem is spoiled by including the answer in the statement of the problem. The problem in the first section furnishes a nice illustration; here is another:

*Example 27*

*Problem A.* Prove that $(n-1)! \equiv 0$ mod 4 if and only if $n$ is a composite $> 4$.

*Problem B.* For which positive integers $n$ is $n$ a factor of the product $(n-1)(n-2)\ldots 3 \cdot 2 \cdot 1$? (For example, 5 *is not* a factor of $4 \cdot 3 \cdot 2 \cdot 1 = 24$, but 6 *is* a factor of $5 \cdot 4 \cdot 3 \cdot 2 \cdot 1 = 120$.)

In the form of Problem A, this example is typical of college texts on number theory; stated in the form of Problem B it could be posed in middle school immediately following the introduction of the concepts *factor* and *prime*. (In fact, it is almost an alternative characterization of "prime.")

Brevity is not necessarily a virtue in posing a problem. Including illustrative examples, for instance, simply enlarges the set of potential solvers.

As a rough rule of thumb, any problem whose statement includes the words "prove that," "show that," and so on, will not encourage guessing and often can be rephrased to do so.

Even problems having only one numerical answer or, perhaps, no answer at all can be posed in this way.

*Example 28:* What values are possible for the area of quadrilateral *EKDL* if *ABCD* and *EFGH* are squares of side 12 and *E* is the center of square *ABCD*?

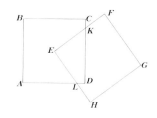

*Example 29:* Find all primes that are one less than a fourth power.

"Find all" is another excellent way to pose an open-search problem that requires guessing or searching for patterns.

Here are several other suggestions for posing good open-search problems:

▶ *"Twenty Questions."* Think of some mathematical object, such as a number, geometric figure, concept, or theorem. (If it is a physical object, you can place it in a bag for added effect.) The students then attempt to identify it by asking questions, as few as possible, which you can answer only with yes or no. Discourage the question "Is it a ____?" Such an activity could be used prior to giving open-search problems as a means to encourage guessing—something that is often hard for teachers to do.

▶ *Whimsical problems.* All too often, students regard textbook problems as artificial; one remedy, cited earlier, is to make the problem as realistic as possible. Another alternative is to pose the problem in a totally unrealistic manner, making it a *whimsical* problem. Such problems often paint a vivid, sometimes ridiculous, picture in the mind. The locker problem mentioned earlier is one example—can't you just see someone walking down a row of 10 000 lockers, opening every other one? Here is another:

*Example 30:* Two pirates were burying their booty on an island, with the Coast Guard in hot pursuit. Near the shore were two large rocks and a lone palm tree. Bluebeard A started from one of the rocks and, walking along the line at right angles to the line joining the rock to the palm tree, paced off a distance equal to that between the rock and the palm tree. Bluebeard B did a similar thing with respect to the other rock and the palm tree. They then walked toward each other and buried the treasure halfway between. Two years later the pirates returned to the island to dig up their treasure but found that the palm tree was no longer there. How can they find the treasure?

A whimsical problem can appeal to students because they do not regard it as phony and it may pique their intellectual curiosity.

Mathematical games and puzzles are still another rich source of open-search problems. They are usually posed as questions and often are whimsical problems as well.

## Re-posing Problems

▶ *Given any skill or concept, it is usually possible to pose a set of nonroutine problems of varying type and difficulty involving that skill or concept.*

In this section we apply some of the foregoing suggestions to the re-posing of problems.

*Example 31 (The straight line):* Typical problems involve finding the equation of a line given two points on the line, graphing a line given its equation, determining the equation of a line given its graph, and so on. Several nonroutine problems follow:

*Problem A*(Recognition). Give the equations of three lines containing the point (1, 2).

*Comment:* The nonspecific answers required should help those students who are merely memorizing the various formulas. Students who choose $x = 1$, $y = 2$, $y = 2x$ might be regarded as clever; they are insightful problem solvers.

*Problem B*(Algorithmic). Take a point, *P*, on the parabola $y = x^2$, $0 \leq x \leq 2$, and compute the slope of the line containing *P* and (1, 1). Do the same for six other points on the curve. What do you think the slope of the line tangent to $y = x^2$ at (1, 1) is? (See fig. 2.)

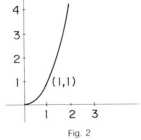

Fig. 2

*Comment:* This is a sequence of algorithmic problems with a purpose.

*Problem C*(Open search). Find all possible values for the slope of a line that contains the origin and intersects the circle $(x - 12)^2 + (y - 5)^2 = 25$.

*Problem D.* Write the equations of several line segments with $-2 \leq y \leq 2$ that will spell your first name.

▶ The problem of posing and re-posing algebraic word problems is a tough one. The solution of word problems requires the translation of English words into mathematical terms—a critically needed skill in any discipline that uses mathematics. One difficulty is that the quantity sought in a word problem requiring the solution of an equation is often known in real life. When the problem is posed realistically, however, it becomes an arithmetic problem.

*Example 32:*

*Problem A.* The Bigtown Coliseum has 20 000 seats—8 000 reserved seats and 12 000 general admission seats. Reserved seats cost $10 each, and general admission seats cost $5 each. The newest rock-group sensation, SMOOCH, is giving a concert there. If $130 000 was collected from the sale of 18 500 tickets, how many reserved seats were sold?

*Comment:* Surely the person selling the tickets knows how many were sold. At best that person would want to compute the amount of money received—an arithmetic problem—if that amount of money were not known.

This might be a somewhat more realistic version of the problem:

*Problem B.* The Bigtown . . . concert there. The promoter estimates her expenses for ushers, programs, janitorial service, and so on, at $10 000. If SMOOCH demands 40% of the price of each ticket sold plus $40 000, how many tickets must she sell to break even? To make a profit of $50 000?

*Note:* Attempts at humor (however feeble) in the statement of mathematical problems are usually worthwhile.

▶ In order to find interesting open-search problems for your class, you may consult one of the many problem books currently available. Offered here, then, are three examples of such problems with suggestions for re-posing them in a more challenging manner:

*Example 33:*

*Problem A.* Prove that a positive integer greater than 9, all of whose digits are identical, cannot be a perfect square.

*Problem B* (Better). Which squares have identical digits? (For example, can 1111 . . . 11 ever be a square?)

*Problem C* (Best). What is the maximum number of identical nonzero digits in which a square can end?

*Comment:* Problems B and C re-pose the problem as a question to encourage guessing. Version C has a positive answer (rather than ''none'').

*Example 34:*

*Problem A.* Show that

$$1 + \frac{1}{2} + \frac{1}{3} + \cdots + \frac{1}{n}$$

is never an integer.

*Problem B.* The series

$$\sum_{k=1}^{\infty} \frac{1}{k}$$

is known to diverge, and consequently the sum

$$\sum_{k=1}^{n} \frac{1}{k} = 1 + \frac{1}{2} + \frac{1}{3} + \cdots + \frac{1}{n}$$

can become arbitrarily large. If we examine a few of its partial sums, we notice that

$$\sum_{k=1}^{4} \frac{1}{k} = 2.083, \quad \sum_{k=1}^{11} \frac{1}{k} = 3.02, \quad \sum_{k=1}^{31} \frac{1}{k} = 4.0272.$$

Can

$$\sum_{k=1}^{n} \frac{1}{k}$$

ever be an integer?

*Comment:* Giving some numerical examples often helps clarify the problem and makes it more tantalizing to the solver.

*Example 35:*

*Problem A. AEDC* and *BCFG* are squares on the legs of right triangle *ABC.* Lines *EB* and *AG* are drawn. Prove that the area of triangle *AHB* equals the area of quadrilateral *CIHJ.*

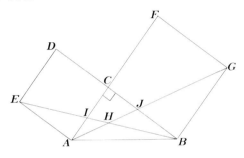

*Problem B. AEDC . . .* drawn. If *BC* = 8 and *AB* = 10, find the areas of triangle *AHB* and quadrilateral *CIHJ.* Can you generalize?

*Comment:* By quantifying the problem, we retain the guessing feature.

## Summary

This article is concerned with the improvement of the oft-neglected art of problem posing. The three key words in this sentence are *neglected*, *art*, and *improvement*. One has only to examine the problem sections of most textbooks to see the "neglect." They often consist of lists of algorithmic exercises, unimaginative word problems, "prove that" problems, and so on. (The review problems are usually worse.) As for "art," it requires the creativity of an artist to pose a problem so that the potential solver will—

1. be motivated to solve the problem;
2. understand and retain the concept involved in the solution of the problem;
3. learn something about the art of solving problems.

The suggestions offered here will, perhaps, improve the state of the art.

The study of mathematics is solving problems. It is incumbent on teachers of mathematics at all levels, therefore, to teach the art of problem solving. The first step in this process is to pose the problem properly.

*5*

# Untangling Clues from Research on Problem Solving

## Marilyn N. Suydam

**A** TANGLED WEB is a phrase that could be applied to problem solving—and to research on problem solving. A real problem is a challenge—as is research on problem solving. Let us apply some techniques we've learned about untangling webs and solving problems to discover what clues research gives us about how to help children become better problem solvers. (These clues are just those that caught my attention; undoubtedly there are others that might have been included.) As writers so often do, I shall digress from the search to examine a bit of background on problem-solving research, for the clues arise from that background.

### The status of the web

Much concern has been expressed all around the country about achievement, especially about achievement in computation. It is increasingly recognized that computation scores are not as poor as scores for the application of computational and other skills in problem solving. Data on problem solving for students aged nine, thirteen, and seventeen from the First National Assessment of Educational Progress were discouraging enough for Carpenter, Coburn, Reys, and Wilson (1976) to state in their interpretive analysis:

> We suspect that solving word problems has not been part of the mathematics program for many 9-year-olds. . . . It is reasonable to believe that the errors were inherent in not knowing what to do with a word problem. [pp. 390–91]

Data from several state assessments (e.g., Michigan and Missouri) further document the point that achievement on problem-solving tests is lower than we (and parents) would like it to be. Carpenter et al. (1976) conclude:

> It is most disturbing to entertain the suggestion that many students receive very little opportunity to learn to solve world problems. The assessment results are so poor, however, that we wonder whether this is not the case. A commitment to working and thinking about word problems is needed for teachers and their students. [p. 392]

We want children to know how to solve mathematical problems: that is one of the primary reasons why mathematics is in the curriculum. Computation, measurement,

34

geometry, algebra, and all the other components of the curriculum are taught because they have applicability in real-life problems.

"Word problems," which come to mind whenever problem solving is discussed,
came into being because educators could find no other plausible way to communicate the applicability of the skills and ideas of mathematics lessons to everyday life.
(There are other means, however.) Teachers recognize the shortcomings of word
problems, just as many deplore the fact that sets of problems often are actually sets
of exercises. The learner doesn't really have to solve a problem—he or she merely
has to read, or just extract information, in order to practice a skill just taught. Robinson (1977) distinguishes between the two in this way:

> An exercise is a verbal rendition of a mathematical sentence, or a verbal description of a
> situation to which some specific process applies. . . . An exercise, by its very definition, is
> specific to a process and is used for practice in that process. . . . Problems requiring crea-
> tivity, or originality . . . are descriptions of situations for which no routine process has been
> previously learned. [p. 22]

Most research on problem solving during this century has focused on word problems, which have frequently been exercises rather than problems—especially at the
elementary school level. During the past decade, however, increasing attention has
been given to problem solving as a process. We have moved from an emphasis on
the characteristics of problems to a consideration of the characteristics of learners.
And even more importance is being attached to the question of just how *do* children
solve problems. As we learn the answers to that question, they may tell us more
about how to teach problem solving. Research has provided some suggestions
about problem solving throughout this century; the findings frequently confirm common sense. Much of this research was conducted at the elementary school level in
response to teachers' expressed concerns. Many of these findings have filtered into
practice through teacher education courses and, more often, through children's textbooks. Suydam and Weaver (1977) culled the research on problem solving at the
elementary school level and presented a list of findings applicable to the classroom.

This article is essentially another culling, an attempt to synthesize research findings across the school curriculum, K–12. It attempts not merely to digest what the
research says but also to call attention to directions for classroom practice that are
indicated by research findings and by researchers who, as a result of their investigations, sought to extrapolate meanings for instructional practice. The intent is to provide teachers at all levels of school mathematics with some suggestions or guidelines
drawn from research, especially research of the past decade. The details of the studies will not be presented; these can be found in the reports of the studies themselves.
Other reviews have also noted details as well as perspectives on the research on
problem solving (e.g., Gorman [1968]; Hatfield [1978]; Kilpatrick [1969]; Suydam
and Weaver [1975]; Trimmer [1974]).

Now let us begin to untangle the web.

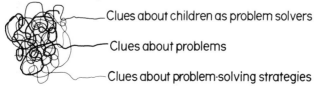

Clues about children as problem solvers

Clues about problems

Clues about problem-solving strategies

First, we shall pick at the loose string tagged "Clues about Children as Problem Solvers." We really do have to begin with the learners—what they're like and how they handle problems. Then we shall tackle "Clues about Problems," the tangible aspect with which we operate. And, finally, we shall unravel "Clues about Problem-solving Strategies," which may help us to plan for more effective teaching of problem solving.

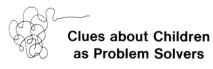

## Clues about Children as Problem Solvers

### Characteristics of good problem solvers

In one sense it does little good to list characteristics of children who are success-ful—or unsuccessful—at solving problems, because we as educators seem to be able to do little about some of them in any direct way. Such lists of characteristics do, however, indicate a few directions in which we can attempt to move all children. As one might expect, good problem solvers tend to have relatively high IQ scores and reasoning ability, high reading comprehension scores, high quantitative ability or computation scores (contributing to success on numerical problems), and high spatial aptitude scores (contributing to success on geometric problems) (Dodson 1971; Heseman 1976; Hollander 1974; Kilpatrick 1968; Moses 1978; Robinson 1973; Talton 1973). It is obvious that to improve your chances of being successful you must select students who are bright, reason well, read well, compute well, and perceive spatial relationships well. Since few teachers can select only this type of child to teach, the task might seem hopeless—but don't give up! You can have some specific influence over other characteristics.

Drawn from several sources (Dodson 1971; Hollander 1974; Krutetskii 1976; Robinson 1973; Suydam and Weaver 1977; Talton 1973), the following composite list includes other characteristics of good problem solvers:

1.  The ability to understand mathematical concepts and terms
2.  The ability to note likenesses, differences, and analogies
3.  The ability to identify critical elements and to select correct procedures and data
4.  The ability to note irrelevant detail
5.  The ability to estimate and analyze
6.  The ability to visualize and interpret quantitative or spatial facts and relation-ships
7.  The ability to generalize on the basis of few examples
8.  The ability to switch methods readily
9.  Higher scores for self-esteem and confidence, with good relationships with other children
10. Lower scores for test anxiety

┌─── **CLUES FOR TEACHING** ─────────────────────────────┐

Let us consider these points one by one and note just one experience that a child could have to promote facility on that point (you can think of many others):

1. Use a term (for instance, *equals*) in a way that makes clear first its mathematical sense and then its use in life outside the school.
2. Classify problems. (Students will probably select "addition" or "baseball" as problem types; when given all "baseball" problems, what other comparisons can be made?)
3. Cite only those aspects of a problem without which there would be no problem.
4. "Cross out" irrelevancies in a word problem, a picture problem, an oral problem, or other form of a problem.
5. Estimate answers and analyze the routes taken to attain the estimates.
6. Describe spatial and numerical ideas not only in words but also by drawing pictures, using physical materials, and building models.
7. Guess what rule or function might apply to a sequence of numbers after exploring a few examples; then verify the guess.
8. Use a variety of methods; this implies that the child knows a variety of methods. (This point is discussed further in a later section.)
9. Praise the child; it still works wonders.
10. Involve the child in using tests as a way for the *child* (and not just the teacher) to evaluate learning.

└──────────────────────────────────────────────────────────┘

Perhaps the most cited researcher in recent years is the Russian, Krutetskii (1976). From continuing investigations of good, capable students, he concluded that in addition to the ability to generalize on the basis of few examples and the ability to switch methods readily (listed above) they skipped steps, had a feeling for elegant solutions (taking the minimum number of steps necessary to solve the problem), and reversed steps easily. Good and poor problem solvers differed in their recall of information from problems. Good problem solvers tended to forget the details of a problem and recall its structural features, whereas poor problem solvers tended to recall the specific details. Recent work in the United States has confirmed these findings (Silver 1979).

┌─── **CLUE FOR TEACHING** ──────────────────────────────┐

How can we help children focus less on details? One idea that may help: Don't focus *all* their attention on those details.

└──────────────────────────────────────────────────────────┘

Positive attitudes toward mathematics were found to be correlated with success in some studies (e.g., Dodson [1971]) and not correlated in others (e.g., Kilpatrick [1968]). Both good and poor problem solvers chose mathematics as the subject least liked, but good problem solvers chose mathematics as their favorite subject more frequently than poor problem solvers did (Robinson 1973).

Lack of concern about messiness or neatness has also been related to success on problem solving (Dodson 1971). Need more be said?

The strategies, or methods, by which pupils approached problems were not consistent for individual pupils from problem to problem, nor were the strategies used similar from problem to problem. Good problem solvers used a formal strategy more often than poor problem solvers, who tended to rely more frequently on a random trial-and-error strategy (Robinson 1973).

---

**CLUE FOR TEACHING**

Teach children a variety of strategies that they can apply in different problem-solving situations, plus an overall plan for how to go about problem solving. More about this later!

---

Good problem solvers took more time to solve novel problems than poor problem solvers did (Kalmykova 1975; Robinson 1973).

---

**CLUE FOR TEACHING**

Allow children sufficient time to solve problems. Not all problems can be solved within the confines of one mathematics lesson; some need to be "mulled over" for longer periods of time. Of course, some problems can be solved relatively quickly, but encourage the expectation that the path to solution will not always be obvious.

---

When boys' and girls' scores on problem-solving tests are compared, there seems to be little difference through most of the elementary school years. However, when significant differences do occur, they are more apt to be in the boys' favor (Fennema 1974). The implications of these findings for instruction are still unclear because we don't know what may cause the difference. To encourage girls to believe that they can solve problems as well as boys would not be amiss, however.

### The general nature of problem solving

Most of us are probably familiar with at least one list of steps indicating how the learner proceeds or should proceed to solve a problem; Polya's (1973) *How to Solve It* is widely known. Each writer brings individual perspectives or philosophies to bear, ranging from pragmatism to the sophistication of artificial intelligence models; moreover, each writer perceives readers differently. A composite of the lists of steps that have been proposed by, for example, Hollowell (1977), Kochen, Badre, and Badre (1976), and LeBlanc (1977) would certainly include these:

1. Understanding the problem—an awareness of the problem situation that stimulates the person to generate a statement of the problem in writing, orally, or merely in thought

2. Planning how to solve the problem

*a*) Break down the components; enumerate data; isolate the unknown

*b*) Recall information from memory; associate salient features with promising solution procedures

*c*) Formulate hypotheses or a general idea of how to proceed

3. Solving the problem

*a*) Transform the problem statement into a mathematical form, or construct representations of the problem situation

*b*) Analyze the statement into subproblems for which the solution is more immediate

*c*) Find a provisional solution

4. Reviewing the problem and the solution

*a*) Check the solution against the problem

*b*) Verify whether the solution is correct; if not, reject the hypotheses, the method of solution, or the provisional solution

*c*) Ascertain an alternative method of solution

The number of possible hypotheses in the second step appears to be a function of the knowledge of relevant information acquired prior to the problem at hand, the type of initial focus as the learner meets the problem, and the parameters of the problem (Miller 1971). This finding has implications for both the selection of problems and the manner of presenting them, as well as the type of questions teachers may need to ask to help the child clarify the problem.

Each of the points on the list needs to be translated into detailed procedures for *how* we can help children to understand, plan, solve, and review. Research has only begun to explore the strategies that children use to solve problems; these strategies in turn provide several clues about strategies that teachers can use.

---

**CLUES FOR TEACHING**

▶Use questions to focus the child's attention on the pertinent information given in the problem.

▶Encourage children to consider different strategies that might be used to solve the problem.

▶After they have reached a solution, encourage children to look back—to reconsider their own thinking, to describe it, to note how they might have solved the problem differently, better, or more efficiently (LeBlanc 1977).

---

The strategy most frequently used to ascertain how children solve problems is the "think aloud" procedure: children are asked to talk about how they are going about solving a problem, and the teacher or investigator asks questions in an attempt to probe the child's mental actions. Underlying these questions may be a model of the nature of problem solving. Many models have been proposed, of which Polya's (1973) set of heuristics is probably the most widely known. Researchers have distinguished between two methods of problem solving—algorithmic and heuristic. The algorithmic method concerns information in the form of rules or operations; it is faster but specific to given types of problems. The heuristic method involves such factors as transformation, goal reduction, and application; it is slower but more general and flexible.

There appears to be a difference in the way students tackle simple and difficult problems:

1. On complex tasks, random routes seem to appear in sequence.
2. On simple tasks, a gestalt seems to operate as an organizing principle for solving problems (Peluffo 1969).

Strategies or heuristics that children use are considered in the section "Clues about Problem Solving."

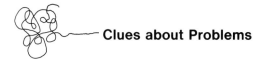

## Clues about Problems

Whoever develops or chooses problems must know something about "good" problems. When textbook authors or teachers develop or select a problem, they must consider (1) its interest level and (2) its difficulty level. What makes a problem interesting to children? Almost any problem can be of interest to *some* children. Although a few studies have investigated what types of problems students prefer, no relationship of these preferences to problem-solving success was found (Cohen 1977; Travers 1967). If textbooks continue to fail to present problems that challenge pupils and that provide teachers with meaningful instructional materials for developing problem-solving skill, then teachers must accept the challenge.

Research has focused, in part, on what makes problems difficult or easy. Many persons (e.g., Cromer [1971]; Jerman [1972]; Loftus [1970]; Meyer [1976]; Rosenthal and Resnick [1971]; Webb [1979]; Whitlock [1974]; and Zalewski [1978]) have studied the difficulty level of problems by using such techniques as factor analysis, in which a computer manipulates data on students' answers until the data cluster into a number of groupings, or by ascertaining the relationship of the data from any one factor to total success. In general, such factors can be classified as mathematical, reading, and reasoning factors; another categorization is by order, digital, and process variables (Cromer 1971).

The findings from these studies have been discrepant, usually because of differences in the types of problems, levels of students, and specific variables being considered. Beardslee and Jerman (1973) explored the difficulty level of a large number of variables and reported that none of them accounted for a significant portion of the variance in difficulty level for the computational problems considered. The studies suggest that factors of mathematical achievement accounted for the most variance (Sagiv 1978; Webb 1979; Zalewski 1978), although, as Meyer (1976) indicated, knowing mathematical skills and concepts does not guarantee that problem solvers will be successful. The studies suggest that a problem is more difficult to solve when—

- it has a lot of words;

- it has more than one step or requires more than one operation for solution;
- the sentence structure is complex;
- it is different in type from the problems preceding it;
- information is not presented in the order in which it will be used;
- it has large numbers.

---

**CLUE FOR TEACHING**

Because these factors make problems difficult does *not* mean that we should avoid such problems. On the contrary, children need experiences with a range of problems containing elements that make them difficult. That, after all, *is* problem solving! We should, however, be wary about giving children problems at a difficulty level too high for their capabilities: we want them to be able to solve most (though not necessarily all) of the problems they meet.

---

Most of us recognize that problems containing extraneous data or information are more difficult and take more time to solve than problems with no extraneous data (Arter and Clinton 1974; Biegen 1972; Blankenship and Lovitt 1976; Fafard 1977).

---

**CLUE FOR TEACHING**

Provide practice on problems having extraneous data. Most real-life problems are like this. Try searching newspapers for items; then pose a question that will demand that extraneous information be disregarded.

---

The largest proportion of correct responses occurs, not surprisingly, when key words act as cues for the procedure or operation that can be used for solution. The proportion of correct answers is much smaller when key words appear as distractors, indicating an incorrect operation or procedure (Nesher 1976; Wright 1968).

---

**CLUE FOR TEACHING**

Children can profit from some instruction directed toward key words, but this instruction should point out the dangers as well as the benefits of relying on key words.

---

When prealgebra students were asked to sort a collection of word problems into groups of mathematically related problems, four dimensions were identified: mathematical structure, contextual details, question form, and pseudostructure (problem similarity based on the presence of a common measurable quantity, such as age or weight) (Silver 1979). The perceived salience of mathematical structure was significantly related to problem-solving competence, confirming the conclusion of Krutetskii (1976) cited earlier.

What is the relationship between reading ability and success in problem solving? At the early primary level, it is obvious that reading can be a deterrent to problem-solving success but, equally obviously, only when problems must be read. There is abundant evidence that young children can solve oral or situational problems prior to formal instruction on problem solving, and all of us have seen instances of it! Moreover, children can use ideas of addition, subtraction, multiplication, and division even though they may not be able to use symbols to represent those ideas (Ibarra and Lindvall 1979; Suydam and Weaver 1975). Counting is, of course, frequently used.

Studies at the upper elementary level do not indicate that reading is as big a deterrent as commonly believed. Some studies, of course, report a positive relationship between reading ability and success in problem solving, but it may not be of sufficient magnitude to be an accurate predictor of problem-solving success (e.g., Harvin and Gilchrist [1970]). In one study, poor reading was not a factor in half of the incorrectly solved standardized test problems; on further analysis, reading difficulties appeared to have accounted for no more than 10 percent of the errors (Knifong and Holtan 1976, 1977). Children who could read the problems simply could not solve them.

Little success has been realized in developing instructional sequences to improve mathematical reading ability (e.g., see Henney [1971]). Perhaps factors other than reading far override the importance of the reading factor.

Manipulative materials, pictures, diagrams, and similar aids enhance the probability of a problem's being understood and solved correctly (Ammon 1973; Bolduc 1970; Caldwell 1978; Grunau 1976; Houtz 1974; Ibarra and Lindvall 1979; Kellerhouse 1975; LeBlanc 1968; McDaniel et al. 1973; Neil 1969; Nelson 1975; Nickel 1971; Portis 1973; Sherrill 1973; Steffe 1967; Steffe and Johnson 1971). Rarely do studies indicate that there is no relationship between the use of materials and problem-solving success (Robinson 1973). The evidence supporting the use of concrete and pictorial aids agrees with that found across mathematics education studies, K–8 (Suydam and Higgins 1977).

---

**CLUE FOR TEACHING**

Have children use, or make available for them to use, manipulative materials, pictures, diagrams, charts, graphs, and so on.

---

Some studies have considered the structure of problems at early grade levels (e.g., Steffe and Johnson [1971]). Suggestions for first-grade teachers that are related to some of these studies can be found in Underhill (1977).

For a number of years, Nelson and others have attempted to structure problems that are "good" for young children who are not sophisticated in the use of mathematical symbols (Bourgeois and Nelson 1977; Bana and Nelson 1977, 1978). The physical structure of a problem can make one more difficult to solve than others, and the "requirements of one task can influence young children's choice of procedure in attempting another task when the same apparatus is involved" (Bourgeois and Nelson 1977, p. 184).

Nelson and his coworkers have determined that many young children seem to experience difficulties in solving problems because of their attention to irrelevant aspects (termed *noise* or *distractors*) of the problem situation. Thus, children classified by such attributes as color, size, or kind when attempting to make equivalent sets.

It appears that distracted children identify and attempt a different problem from the one assigned. . . . For the distracted child the relevant information becomes irrelevant. . . . It appears that a distraction only comes into play if it forms the basis for some plausible alternative problem for the child. [Bana and Nelson 1977, pp. 276–77]

About one-third of the children used manipulatives, whereas others gave verbal responses without any manipulations (Bana and Nelson 1977). Most of the children who used an invalid process based their solutions on irrelevant spatial-numerical cues.

---

**CLUE FOR TEACHING**

"Different settings can yield different performance. If a child succeeds in one setting teachers should not be deluded into thinking that the task has been mastered. Children should be subjected to the same problem in a variety of situations. . . . Children need to develop the ability to cut through noise in order to abstract mathematics from the environment. The findings suggest that young children [grades 1–3] are not being given sufficient experience with manipulative materials or with partitive division problems, as indicated by the reluctance of many subjects to manipulate materials and the lack of systematic approaches. Finally, the low success rate of impulsive responders suggests that teachers should not hurry children's responses; rather, they should encourage children to reflect on each problem and thus provide more opportunity to cope with distractions" (Bana and Nelson 1977, p. 278).

---

# Clues about
# Problem-solving Strategies

Teaching a general strategy or heuristic versus teaching specific strategies or heuristics has been the focus of much discussion over the years. (Incidentally, arguments over terminology also abound.) Common sense indicates that both types of strategies need to be taught. Children need an overall procedure for attacking a problem; this provides a global sense of security. Then they need specific strategies that they can apply within the global structure.

Research evidence strongly concurs that problem-solving performance is enhanced by teaching students to use a variety of strategies or heuristics, both general and specific. That is, students using a wide range of strategies are able to solve more problems (Blake 1977; Graham 1978; Pennington 1970; Webb 1979; Wilson 1968). Moreover:

- When heuristics are specifically taught, they are then used more and students achieve correct solutions more frequently (Clement 1977; Fowler 1978; Hall 1976; Kantowski 1977; Lee, J. 1978; Lee, K. 1978; Vos 1976).

- Training in a variety of heuristics is necessary so that students have a repertoire from which they can draw as they meet the wide variety of problems that exist; different mathematical content evokes different strategies. Certain strategies (e.g., analysis and synthesis) are used more frequently than others (at particular ages) (Brandau and Dossey 1979).

- Various strategies are used at different stages in solving problems. Thus, training on *both* integration (the capacity to integrate remaining components in a sequence of operations required for problem solution) and evaluation (capacity to judge whether the solution is correct) seems to be required for solving problems; when either was absent, the solution rate did not exceed chance (Gallo 1975).

---

**CLUE FOR TEACHING**

Flexibility in problem solving is a type of learned behavior. Students who are exposed to a variety of problems are able to make a smoother transition to new problems than those who are given practice only on many similar problems. Questioning may also contribute to the development of flexibility in children's problem-solving behaviors (Cunningham 1966).

---

- The use of regular patterns of analysis and synthesis was noted in the solutions of higher-scoring students. Many of these regular patterns were immediately preceded by a goal-oriented heuristic. Students who had no direction tended to establish as many facts as possible whether they were necessary for the solution or not (Kantowski 1977).
- Deduction and trial-and-error patterns were also found (Dalton 1975).
- A sequence in which problem types were moderately varied was better than either a highly varied sequence or a nonvaried sequence (Sumagaysay 1972).

---

**CLUE FOR TEACHING**

As one example related to this point, Smith (1973) suggests that problems involving related theorems should be alternated so that students are made aware of the essential elements of a theorem and do not develop a "set" to use a particular theorem when it is not applicable.

---

- The number of times a student attempted to solve a problem was unrelated to obtaining a correct solution. Changing the mode of attack in solving a problem was, however, significantly related to obtaining a correct solution (Blake 1977).
- Students at different developmental levels tend to differ in the extent to which they use particular strategies; for instance, students at a formal-operational level (as defined by Piaget) used more means-end heuristics than concrete-operational students did (Grady 1976). Formal-operational students also used a larger variety of heuristics (Days 1978). They used deduction, evaluation, and systematic trial-and-error strategies on significantly more problems.
- Similarly, the level of conservation was related to problem-solving achievement

at younger age levels (LeBlanc 1968; Shores and Underhill 1976; Steffe 1967).

- Students who scored high on the divergent type of problem made fewer generalizations and used trial-and-error strategies more often (Maxwell 1975).

- Logical analysis was widely used by fourth graders; creative or divergent thinking was most successful but seldom used. Blind guessing and nonsystematic trial and error were the most unsuccessful strategies (Sanders 1973).

- A common strategy for fifth graders seemed to be to modify, rather than abandon, the most strongly cued procedure when its inapplicability was apparent (Morris 1976).

- Problem-solving skills are improved by incorporating them throughout the curriculum—that is, organizing the curriculum as a sequence of problems in which students induce organizational rules from examples (Roman 1975).

---

**CLUES FOR TEACHING**

Several general clues are provided by Flener (1978, p. 13):

▶ "Whenever possible, embed a teaching/learning experience in a problem-solving format.

▶ Think in terms of hints or suggestions rather than absolute procedures to be followed.

▶ Experiment by giving students less help than usual.

▶ Do not be misled by the immediate benefits of structural [expository] teaching—the long-range benefits of teaching through problem solving may be higher."

---

There is no one optimal strategy or heuristic for problem solving. In his review of the literature that focused on the relationships among instructional method, internal cognitive activity, and performance measures, Mayer (1974) concluded that little progress will be made until the emphasis on "which method is best" gives way to an attempt to define and relate to one another (1) the external features of instruction, (2) the internal features of subjects' characteristics, (3) the activity during learning, and (4) the outcome performance measures. A start has been made as researchers delve into processes; much more needs to be learned.

---

**CLUES FOR TEACHERS**

"It appears that techniques which attempt to 'program' the solver to follow a fixed sequence of steps are not very effective. . . . Experience and some research suggest, however, that certain heuristic procedures which will improve mathematical problem-solving performance can be learned—provided the teacher illustrates how the procedures work; gives ample opportunity for discussion, practice, and reflection; and supports and encourages the learner's efforts." (Kilpatrick 1978, p. 191)

---

## Concluding Clues

If we present problems to children that challenge them to think and if we develop the idea that they must search for a solution, we shall have taken a big step toward promoting better problem solving at all levels.

> Students need to be faced with problems in which the approach is not apparent and encouraged to generate many alternative approaches, testing each hypothesis generated. In this way students will come to understand the nature of problem solving. [Wheatley 1977, p. 38]

If they learn techniques to aid them in the search, they should then be able to solve even the problems on current standardized tests without relying on instant recognition of which algorithm or procedure they should use.

Research from brain-hemisphere processing, information processing, and other sources has potential implications for mathematics education. Within mathematics education, research must continue trying to ascertain the ways in which different types of learners solve different types of problems with different types of content; furthermore, appropriate instructional materials must continue to be produced.

But the real clue lies with teachers: only they can apply the clues and use the materials to help children untangle the web of problem solving.

---
**CLUES FOR TEACHING**

Over a quarter of a century ago, Brownell (1942) developed a list of suggestions for teachers. Three of them seem particularly appropriate in this section on concluding clues:

- "To be most fruitful, practice in problem solving should not consist in repeated experiences in solving the same problems with the same techniques, but should consist of the solution of different problems by the same techniques and the application of different techniques to the same problems" (p. 439).

- "A problem is not necessarily 'solved' because the correct response has been made. A problem is not truly solved unless the learner understands what he [or she] has done and knows why his [or her] actions were appropriate" (p. 439).

- "Instead of being 'protected' from error, the child should many times be exposed to error and be encouraged to detect and to demonstrate what is wrong, and why" (p. 440).

---

### REFERENCES

Ammon, Richard I., Jr. "An Analysis of Oral and Written Responses in Developing Mathematical Problems through Pictorial and Written Stimuli." (Doctoral dissertation, Pennsylvania State University, 1972.) *Dissertation Abstracts International* 34A (1973): 1056–57.

Arter, Judith A., and LeRoy Clinton. "Time and Error Consequences of Irrelevant Data and Question Placement in Arithmetic Word Problems II: Fourth Graders." *Journal of Educational Research* 68 (September 1974): 28–31.

Bana, J. P., and Doyal Nelson. "Some Effects of Distractions in Nonverbal Mathematical Problems." *Alberta Journal of Educational Research* 23 (December 1977): 268–79.

————. "Distractors in Nonverbal Mathematical Problems." *Journal for Research in Mathematics Education* 9 (January 1978): 55–61.

Beardslee, Edward C., and Max E. Jerman. *Linguistic Variables in Verbal Arithmetic Problems.* 1973. (ERIC no. ED 073 926)

Biegen, David A. "The Effects of Irrelevant and Immaterial Data on Problem Difficulty." (Doctoral dissertation, University of Cincinnati, 1971.) *Dissertation Abstracts International* 32A (1972): 3774.

Blake, Rick N. "The Effect of Problem Context upon the Problem Solving Processes Used by Field Dependent and Independent Students: A Clinical Study." (Doctoral dissertation, University of British Columbia, 1976.) *Dissertation Abstracts International* 37A (1977): 4191–92.

Blankenship, Coleen S., and Thomas C. Lovitt. "Story Problems: Merely Confusing or Downright Befuddling?" *Journal for Research in Mathematics Education* 7 (November 1976): 290–98.

Bolduc, Elroy J., Jr. "A Factorial Study of the Effects of Three Variables on the Ability of First-Grade Children to Solve Arithmetic Addition Problems." (Doctoral dissertation, University of Tennesse, 1969.) *Dissertation Abstracts International* 30A (1970): 3358.

Bourgeois, Roger, and Doyal Nelson. "Young Children's Behavior in Solving Division Problems." *Alberta Journal of Educational Research* 23 (September 1977): 178–85.

Brandau, Linda I., and John A. Dossey. "Analyzing Cognitive Behaviors for Conjecturing in Mathematical Problem Solving: An Exploratory Study." In *NCTM Annual Meeting Research Reports.* Columbus, Ohio: ERIC/SMEAC, 1979.

Brownell, William A. "Problem Solving." In *The Psychology of Learning.* Forty-first Yearbook of the National Society for the Study of Education, pt. 2, pp. 415–43. Chicago: The Society, 1942.

Caldwell, Janet H. "The Effects of Abstract and Hypothetical Factors on Word Problem Difficulty in School Mathematics." (Doctoral dissertation, University of Pennsylvania, 1977.) *Dissertation Abstracts International* 38A (1978): 4637.

Carpenter, Thomas, Terrence G. Coburn, Robert E. Reys, and James W. Wilson. "Notes from National Assessment: Word Problems." *Arithmetic Teacher* 23 (May 1976): 389–93.

Clement, John J. "Quantitative Problem Solving Processes in Children." (Doctoral dissertation, University of Massachusetts, 1977.) *Dissertation Abstracts International* 38A (1977): 1952–53.

Cohen, Martin P. "Interest and Its Relationship to Problem-solving Ability among Secondary School Mathematics Students." (Doctoral dissertation, University of Texas at Austin, 1976.) *Dissertation Abstracts International* 37A (1977): 4929.

Cromer, Fred E. "Structural Models for Predicting the Difficulty of Multiplication Problems." (Doctoral dissertation, George Peabody College for Teachers, 1971.) *Dissertation Abstracts International* 32 (1971): 1974.

Cunningham, John D. *Rigidity in Children's Problem Solving.* (ERIC no. ED 010 996) Reprint from *School Science and Mathematics* 66 (April 1966): 377–89.

Dalton, Roy M. "Thinking Patterns in Solving Certain Word Problems by Ninth Grade General Mathematics Students: An Exploratory Study in Problem Solving." (Doctoral dissertation, University of Tennessee, 1974.) *Dissertation Abstracts International* 35B (1975): 5526.

Days, Harold C. "The Effect of Problem Structure on the Processes Used by Concrete- and Formal-Operational Students to Solve Verbal Mathematics Problems." (Doctoral dissertation, Purdue University, 1977.) *Dissertation Abstracts International* 38A (1978): 6038.

Dodson, Joseph W. "Characteristics of Successful Insightful Problem Solvers." (Doctoral dissertation, University of Georgia, 1970.) *Dissertation Abstracts International* 31A (1971): 5928.

Fafard, Mary-Beth. "The Effects of Instructions on Verbal Problem Solving in Learning Disabled Children." (Doctoral dissertation, University of Oregon, 1976.) *Dissertation Abstracts International* 37A (1977): 5741–42.

Fennema, Elizabeth. "Mathematics Learning and the Sexes: A Review." *Journal for Research in Mathematics Education* 5 (May 1974): 126–39.

Flener, Frederick O. "Reflections on a Problem Solving Study." *International Journal of Mathematical Education in Science and Technology* 9 (February 1978): 9–13.

Fowler, Evelyn C. "A Study Interrelating Situational Problem Solving, Mathematical Model Building, and Divergent Thinking among Gifted Secondary Mathematics Students." (Doctoral dissertation, Georgia State University, 1978.) *Dissertation Abstracts International* 39A (1978): 2111–12.

Gallo, Dennis P. "Problem Solving: A Test of Two Aspects." (Doctoral dissertation, State University of New York at Stony Brook, 1974.) *Dissertation Abstracts International* 35B (1975): 4649–50.

Gorman, Charles J. "A Critical Analysis of Research on Written Problems in Elementary School Mathematics." (Doctoral dissertation, University of Pittsburgh, 1967.) *Dissertation Abstracts* 28A (1968): 4818–19.

Grady, Merle B. "Problem Solving in Algebra as Related to Piagetian Levels of Thought." (Doctoral dissertation, University of Texas at Austin, 1975.) *Dissertation Abstracts International* 36A (1976): 6587.

Graham, Verdell G. "The Effect of Incorporating Sequential Steps and Pupil-constructed Problems on Performance and Attitude in Solving One-Step Verbal Problems Involving Whole Numbers." (Doctoral dissertation, Catholic University of America, 1978.) *Dissertation Abstracts International* 39A (1978): 2028.

Grunau, Ruth V. E. "Effects of Elaborative Prompt Condition and Developmental Level on Performance of Addition Problems by Kindergarten Children." (Doctoral dissertation, University of British Columbia, 1975.) *Dissertation Abstracts International* 36A (1976): 4349.

Hall, Thomas R. "A Study of Situational Problem Solving by Gifted High School Mathematics Students." (Doctoral dissertation, Georgia State University, 1976.) *Dissertation Abstracts International* 37A (1976): 906–7.

Harvin, Virginia R., and Mary A. Gilchrist. *Mathematics Teacher—a Reading Reacher?* 1970. (ERIC no. ED 041 702)

Hatfield, Larry L. "Heuristical Emphases in the Instruction of Mathematical Problem Solving: Rationales and Research." In *Mathematical Problem Solving*, edited by Larry L. Hatfield. Columbus, Ohio: ERIC/SMEAC, 1978.

Henney, Maribeth. "Improving Mathematics Verbal Problem Solving Ability through Reading Instruction." *Arithmetic Teacher* 18 (April 1971): 223–29.

Heseman, John P. "A Spatial Model for the Cognitive Representation of Verbal Algebra Problems." (Doctoral dissertation, Indiana University, 1976.) *Dissertation Abstracts International* 37A (1976): 2037.

Hollander, Sheila K. "Strategies of Selected Sixth Graders Reading and Working Verbal Arithmetic Problems." (Doctoral dissertation, Hofstra University, 1973.) *Dissertation Abstracts International* 34A (1974): 6258–59.

Hollowell, Kathleen A. "A Flow Chart Model of Cognitive Processes in Mathematical Problem Solving." (Doctoral dissertation, Boston University School of Education, 1977.) *Dissertation Abstracts International* 37A (1977): 7666–67.

Houtz, John C. "Problem-solving Ability of Advantaged and Disadvantaged Elementary School Children with Concrete and Abstract Item Representations." (Doctoral dissertation, Purdue University, 1973.) *Dissertation Abstracts International* 34A (1974): 5717.

Ibarra, Cheryl G., and C. Mauritz Lindvall. "An Investigation of Factors Associated with Children's Comprehension of Simple Study Problems Involving Addition and Subtraction Prior to Formal Instruction on These Operations." In *NCTM Annual Meeting Research Reports.* Columbus, Ohio: ERIC/SMEAC, 1979.

Jerman, Max E. *Predicting the Relative Difficulty of Problem-solving Exercises in Arithmetic.* Final report, Grant No. OEG 3-72-0036. December 1972. (ERIC no. ED 070 678)

Kalmykova, Z. I. *Analysis and Synthesis as Problem-solving Methods,* edited by Mary G. Kantowski. Soviet Studies in the Psychology of Learning and Teaching Mathematics, edited by Jeremy Kilpatrick, Izaak Wirszup, Edward G. Begle, and James W. Wilson, vol. 11. Chicago: University of Chicago, 1975. (Available from the National Council of Teachers of Mathematics, 1906 Association Dr., Reston, VA 22091.)

Kantowski, Mary G. "Processes Involved in Mathematical Problem Solving." *Journal for Research in Mathematics Education* 8 (May 1977): 163–80.

Kellerhouse, Kenneth D., Jr. "The Effects of Two Variables on the Problem Solving Abilities of First Grade and Second Grade Children." (Doctoral dissertation, Indiana University, 1974.) *Dissertation Abstracts International* 35A (1975): 5781.

Kilpatrick, Jeremy. "Analyzing the Solution of Word Problems in Mathematics: An Exploratory Study." (Doctoral dissertation, Stanford University, 1967.) *Dissertation Abstracts* 28A (1968): 4380.

———. "Problem Solving in Mathematics." *Review of Educational Research* 39 (October 1969): 523–34.

———. "Research on Problem Solving in Mathematics." *School Science and Mathematics* 78 (March 1978): 189–92.

Knifong, J. Dan, and Boyd Holtan. "An Analysis of Children's Written Solutions to Word Problems." *Journal for Research in Mathematics Education* 7 (March 1976): 106–12.

———. "A Search for Reading Difficulties among Erred Word Problems." *Journal for Research in Mathematics Education* 8 (May 1977): 227–30.

Kochen, Manfred, Albert N. Badre, and Barbara Badre. "On Recognizing and Formulating Mathematical Problems." *Instructional Science* 5 (April 1976): 115–31.

Krutetskii, V. A. *The Psychology of Mathematical Abilities in Schoolchildren.* Edited by Jeremy Kilpatrick and Izaak Wirszup and translated by Joan Teller. Chicago: University of Chicago Press, 1976.

LeBlanc, John F. "The Performances of First Grade Children in Four Levels of Conservation of Numerousness and Three I.Q. Groups When Solving Arithmetic Subtraction Problems." (Doctoral dissertation, University of Wisconsin, 1968.) *Dissertation Abstracts* 29A (1968): 67.

————. "You Can Teach Problem Solving." *Arithmetic Teacher* 25 (November 1977): 16–20.

Lee, Jong S. "The Effects of Process Behaviors on Problem-solving Performance in Various Tests." (Doctoral dissertation, University of Chicago, 1978.) *Dissertation Abstracts International* 39A (1978): 2149.

Lee, Kil S. "An Exploratory Study of Fourth Graders' Heuristic Problem Solving Behavior." (Doctoral dissertation, University of Georgia, 1977.) *Dissertation Abstracts International* 38A (1978): 4004.

Loftus, Elizabeth J. F. *An Analysis of the Structural Variables That Determine Problem-solving Difficulty on a Computer-based Teletype.* Report to the National Science Foundation. 1970. (ERIC no. ED 047 505).

Mayer, Richard E. *Learning to Solve Problems: Role of Instructional Method and Learner Activity.* Report to the National Science Foundation. 1974. (ERIC no. ED 091 225)

Maxwell, Ann A. "An Exploratory Study of Secondary School Geometry Students: Problem Solving Related to Convergent-Divergent Productivity." (Doctoral dissertation, University of Tennessee, 1974.) *Dissertation Abstracts International* 35A (1975): 4987.

McDaniel, Ernest, John F. Feldhusen, Grayson Wheatley, and John C. Houtz. *Measurement of Concept Formation and Problem Solving in Disadvantaged Elementary School Children.* Final report, Grant No. OEG-5-72-0023(509). 1973. (ERIC no. ED 090 318)

Meyer, Ruth A. "A Study of the Relationship of Mathematical Problem Solving Performance and Intellectual Abilities of Fourth-Grade Children." (Doctoral dissertation, University of Wisconsin—Madison, 1975.) *Dissertation Abstracts International* 37A (1976): 123–24.

Miller, Lance A. "Hypothesis Analysis of Conjunctive Concept-Learning Situations." *Psychological Review* 78 (May 1971): 262–71.

Morris, Lynn L. "An Information Processing Examination of Skill Assembly in Problem Solving and Development Using Wertheimer's Area of a Parallelogram Problem." (Doctoral dissertation, University of Pittsburgh, 1975.) *Dissertation Abstracts International* 36A (1976): 7957.

Moses, Barbara E. "The Nature of Spatial Ability and Its Relationship to Mathematical Problem Solving." (Doctoral dissertation, Indiana University, 1977.) *Dissertation Abstracts International* 38A (1978): 4640.

Neil, Marilyn S. "A Study of the Performance of Third Grade Children on Two Types of Verbal Arithmetic Problems." (Doctoral dissertation, University of Alabama, 1968.) *Dissertation Abstracts* 29A (1969): 3337.

Nelson, Glenn T. "The Effects of Diagram Drawing and Translation on Pupils' Mathematics Problem-solving Performance." (Doctoral dissertation, University of Iowa, 1974.) *Dissertation Abstracts International* 35A (1975): 4149.

Nesher, Pearla. "Three Determinants of Difficulty in Verbal Arithmetic Problems." *Educational Studies in Mathematics* 7 (December 1976): 369–88.

Nickel, Anton P. "A Multi-Experience Approach to Conceptualization for the Purpose of Improvement of Verbal Problem Solving in Arithmetic." (Doctoral dissertation, University of Oregon, 1971.) *Dissertation Abstracts International* 32A (1971): 2917–18.

Peluffo, N. "Constructions of Relations and Classes and Their Use in Problem Solving." *Rivista di Psicologia Sociale e Archivio Italiano di Psicologia Generale e del Lavoro* 36 (1969): 109–42.

Pennington, Barbara A. "Behavioral and Conceptual Strategies as Decision Models for Solving Problems." (Doctoral dissertation, University of California, Los Angeles, 1970.) *Dissertation Abstracts International* 31A (1970): 1630–31.

Polya, G. *How to Solve It.* 3d ed. Princeton, N.J.: Princeton University Press, 1973.

Portis, Theodore R. "An Analysis of the Performances of Fourth, Fifth and Sixth Grade Students on Problems Involving Proportions, Three Levels of Aids and Three I.Q. Levels." (Doctoral dissertation, Indiana University, 1972.) *Dissertation Abstracts International* 33A (1973): 5981–82.

Robinson, Edith. "On the Uniqueness of Problems in Mathematics." *Arithmetic Teacher* 25 (November 1977): 22–26.

Robinson, Mary L. "An Investigation of Problem Solving Behavior and Cognitive and Affective Characteristics of Good and Poor Problem Solvers in Sixth Grade Mathematics." (Doctoral dissertation, State University of New York at Buffalo, 1973.) *Dissertation Abstracts International* 33A (1973): 5620.

Roman, Richard A. *The Word Problem Program: Summative Evaluation.* Report to NIE and NSF. 1975. (ERIC no. ED 113 212)

Rosenthal, Daniel J. A., and Lauren B. Resnick. *The Sequence of Information in Arithmetic Word Problems.* Paper presented at AERA meeting. 1971. (ERIC no. ED 049 909)

Sagiv, Abraham. "Hierarchical Structure of Learner Abilities in Verbal Computation Problem-Solving Related to Strength of Material." *International Journal of Mathematical Education in Science and Technology* 9 (1978): 451–56.

Sanders, Violet A. "Arithmetic Problem Solving Strategies of Fourth Grade Children." (Doctoral dissertation, Wayne State University, 1972.) *Dissertation Abstracts International* 33A (1973): 5983–84.

Sherrill, James M. "The Effects of Different Presentations of Mathematical Word Problems upon the Achievement of Tenth Grade Students." *School Science and Mathematics* 73 (April 1973): 277–82.

Shores, Jay H., and Robert G. Underhill. *An Analysis of Kindergarten and First Grade Children's Addition and Subtraction Problem Solving Modeling and Accuracy.* Paper presented at AERA meeting. 1976. (ERIC no. ED 121 626)

Silver, Edward. "Problem-solving Competence and Memory for Mathematical Problems: Cue Salience and Recall." In *NCTM Annual Meeting Research Reports.* Columbus, Ohio: ERIC/SMEAC, 1979.

Smith, James P. "The Effect of General versus Specific Heuristics in Mathematical Problem-solving Tasks." (Doctoral dissertation, Columbia University, 1973.) *Dissertation Abstracts International* 34A (1973):2400.

Steffe, Leslie P. "The Performance of First Grade Children in Four Levels of Conservation of Numerousness and Three I.Q. Groups When Solving Arithmetic Addition Problems." (Doctoral dissertation, University of Wisconsin, 1966.) *Dissertation Abstracts* 28A (1967):885–86.

Steffe, Leslie P., and David C. Johnson. "Problem-solving Performances of First-Grade Children." *Journal for Research in Mathematics Education* 2 (January 1971):50–64.

Sumagaysay, Lourdes S. "The Effects of Varying Practice Exercises and Relating Methods of Solution in Mathematics Problem Solving." (Doctoral dissertation, University of Toronto, 1970.) *Dissertation Abstracts International* 32A (1972):6751.

Suydam, Marilyn N., and Jon L. Higgins. *Activity-based Learning in Elementary School Mathematics: Recommendations from Research.* Columbus, Ohio: ERIC/SMEAC, 1977.

Suydam, Marilyn N., and J. Fred Weaver. *Using Research: A Key to Elementary School Mathematics.* Columbus, Ohio: ERIC/SMEAC, 1975.

———."Research on Problem Solving: Implications for Elementary School Classrooms." *Arithmetic Teacher* 25 (November 1977):40–42.

Talton, Carolyn F. "An Investigation of Selected Mental, Mathematical, Reading, and Personality Assessments as Predictors of High Achievers in Sixth Grade Mathematical Verbal Problem Solving." (Doctoral dissertation, Northwestern State University of Louisiana, 1973.) *Dissertation Abstracts International* 34A (1973):1008–9.

Travers, Kenneth J. "A Test of Pupil Preference for Problem-solving Situations in Junior High School Mathematics." *Journal of Experimental Education* 35 (Summer 1967):9–18.

Trimmer, Ronald G. *A Review of the Research Relating Problem Solving and Mathematics Achievement to Psychological Variables and Relating These Variables to Methods Involving or Compatible with Self-Correcting Manipulative Mathematics Materials.* 1974. (ERIC no. ED 092 402)

Underhill, Robert G. "Teaching Word Problems to First Graders." *Arithmetic Teacher* 25 (November 1977):54–56.

Vos, Kenneth E. "The Effects of Three Instructional Strategies on Problem-solving Behaviors in Secondary School Mathematics." *Journal for Research in Mathematics Education* 7 (November 1976):264–75.

Webb, Norman L. "Processes, Conceptual Knowledge, and Mathematical Problem-solving Ability." *Journal for Research in Mathematics Education* 10 (March 1979):83–93.

Wheatley, Grayson H. "The Right Hemisphere's Role in Problem Solving." *Arithmetic Teacher* 25 (November 1977):36–39.

Whitlock, Prentice E. "An Investigation of Selected Factors That Affect Ability to Solve Verbal Mathematical Problems at the Primary Level." (Doctoral dissertation, Fordham University, 1974.) *Dissertation Abstracts International* 35A (1974):1437.

Wilson, James W. "Generality of Heuristics as an Instructional Variable." (Doctoral dissertation, Stanford University, 1967.) *Dissertation Abstracts* 28A (1968):2575.

Wright, Jone P. "A Study of Children's Performance on Verbally Stated Problems Containing Word Clues and Omitting Them." (Doctoral dissertation, University of Alabama, 1968.) *Dissertation Abstracts* 29B (1968):1770.

Zalewski, Claire J. "An Investigation of Selected Factors Contributing to Success in Solving Mathematical Word Problems." (Doctoral dissertation, Boston University School of Education, 1978.) *Dissertation Abstracts International* 39A (1978):2804.

# 6

# Opinions about Problem Solving in the Curriculum for the 1980s: A Report

**Alan Osborne**
**Margaret B. Kasten**

PROBLEM solving is the primary focus of NCTM's curriculum recommendations for the decade of the 1980s. Each of the specific recommendations about basic skills, the emphasis in content and instruction, the uses of calculators and computers, the design of courses and curricula, as well as a number of the proposals for implementing the recommended changes, is directed toward helping students acquire power in problem solving.

The preparation of these recommendations required two years during the late seventies, with a major portion of the first year being given to collecting information from a variety of sources about features desired for the mathematics curriculum. One primary source of information was the NCTM Priorities in School Mathematics (PRISM) Project. (Other members of the project, besides ourselves, were F. Joe Crosswhite, Jon L. Higgins, Patricia Newcomb, and Marilyn N. Suydam.)

Professionals and laypersons were asked in two rounds of surveys about their curricular preferences and priorities for the 1980s. Problem solving was one of nine curricular strands treated in the surveys. The other strands were whole numbers, algebra, geometry, ratio-proportion-percent, probability-statistics, computer literacy, fractions-decimals, and measurement. This report focuses on the major conclusions and observations about the data directly concerned with problem solving. (Descriptions of the specific processes used and the data supporting these conclusions, as well as findings for the other curricular strands, can be found in the reports of the NCTM PRISM Project.)

Two rounds of surveys were conducted; the first concerned preferences, the second, priorities. The majority of specific information concerning problem solving was

This report is based on work supported by the National Science Foundation under NSF Grant No. SED77-18564. Any opinions, findings, and conclusions expressed here are those of the authors and do not necessarily reflect the views of the National Science Foundation.

collected in the first-round preference surveys. Nine population samples received the preference surveys. The conclusions reported below include the opinions of the six samples designated professional:

AT—a sample identified from subscribers to the *Arithmetic Teacher* who do not subscribe to the *Mathematics Teacher* and who are not on the NCTM list of supervisors or teacher educators

MT—a sample identified from subscribers to the *Mathematics Teacher* who do not subscribe to the *Arithmetic Teacher* and who are not on the NCTM list of supervisors or teacher educators

JC—a sample identified from the membership roster of the American Mathematical Association of Two-Year Colleges

MA—a sample identified from the membership of the Mathematical Association of America

TE—a sample selected from the NCTM list of teacher educators

SP—a sample selected from the NCTM list of mathematics supervisors

Each sample was selected on a random basis. Techniques of item sampling were employed in order to provide a larger pool of items. Item sampling means, however, that not all individuals responded to the same set of items.

Three samples drawn from outside the mathematics community yielded insights into views of the curriculum held by the lay population. Samples of principals (PR), school board presidents (SB), and presidents of parent-teacher associations (PT) were included. Although there was considerable overlap between the item pools for the lay and the professional groups, the lay group did respond to many questions that were less specific and technical.

The six professional groups responded to a pool of seventy items directly concerned with problem solving. These items were grouped in six clusters corresponding to goals, content, resources, methods, specific questions about types of students and timing of instruction, and the use of the calculator in teaching problem solving. The remaining three samples received eleven items specifically related to problem solving.

Each instrument began with a short set of items designed to orient respondents to thinking about the future before they responded to clusters of items in the nine curricular strands. Individuals in all samples were asked to indicate whether in their judgment a given curricular topic, problem, or issue should receive *substantially more, more, about the same, less,* or *substantially less* emphasis during the coming decade. Forty-five different topics, issues, and problems—such as basic skills, decimals, fractions, algebra, the talented, metric measure, daily homework, and special mathematics for ethnic students—appeared on the instrument forms. When average responses to individual topics were compared, an inferred ranking of these forty-five items was obtained. Across the six professional population samples, problem solving ranked in the top two choices. The AT, TE, JC, and SP populations ranked problem solving first, and the MT and MA samples ranked it second.

The three samples drawn from outside the mathematics education community responded to a pool of only twenty of these topics, issues, and problems. The SB sample ranked problem solving second, paralleling the professional samples; however, the PR and PT samples ranked problem solving fifth. The results of the latter two samples may indicate that these populations view other of the areas listed as more pressing than problem solving. However, in no way should it be viewed as an indication that these samples do not support increasing the emphasis on problem solving. In fact, 84 percent of the PR sample and 76 percent of the PT sample favored increasing the emphasis on problem solving.

The PR, SB, and PT samples also responded to a cluster of items concerning twelve general goals for teaching mathematics. The goal receiving the strongest support across the three samples was *To solve problems in everyday life.* This goal had the highest inferred rank, and at least 98 percent of each of these samples felt it was a very important or somewhat important purpose for teaching mathematics. A further indication of the importance of problem solving to the PR, SB, and PT samples is seen in the positive response given to the following relatively controversial item: *Problems should be realistic even though they might involve sensitive social issues.* The PR, SB, and PT samples supported this statement at the 74 percent, 69 percent, and 66 percent levels, respectively.

For all nine populations, strong support is evident for—

CONCLUSION 1. *Problem solving should receive more emphasis in the school mathematics program during the coming decade.*

Insights into the reasons for emphasizing problem solving in the curriculum were obtained from a cluster of ten items concerned with ten specific, commonly accepted goals for instruction on problem solving. Each goal was described as the featured characteristic of a set of instructional materials. The six professional samples indicated their preferences for the goals by indicating whether they would buy or use the instructional materials for a given goal, even at the expense of accomplishing other goals. Table 1 displays the rank order of the ten goals for each population.

The Spearman rank-order correlation coefficients for pairs of the populations indicate that the following conclusion is warranted:

CONCLUSION 2. *There is broadly based unanimity for the aims or purposes of instruction on problem solving.*

Some differences in responses among populations are notable. The AT, MT, and JC samples are more concerned than the other populations about the role of problem solving in providing skills necessary for living in today's world (item 506). The JC and MA samples (with the closest ties to educating scientists) prize the goal of applying mathematics in science (item 510) more than the other populations.

When the means and mechanisms for realizing the goals associated with problem solving are considered, the discrepancies among populations and the ambivalence within the populations become evident. The fact of the matter is that problem solving has never been widely considered in traditional curricular design terms. For other curricular domains in mathematics, questions and issues are conceived and discussed in terms of scope-and-sequence decisions. For problem solving, we have seldom considered the questions of what specific behaviors are appropriate for a given level of maturity of the learner or what order of experiences in problem solving most facilitate learning. As you examine the data that follow concerning content selection, methods, or resources, remember that what we are teaching when we teach problem solving is not as refined by tradition or experience as the content associated with other areas, such as whole-number computational skills, fractions, geometric

TABLE 1
Rank Orders of Goals for Instruction in Problem Solving

| Goal Statement | population | | | | | |
|---|---|---|---|---|---|---|
| | AT | MT | JC | MA | TE | SP |
| Problem solving is taught— | | | | | | |
| 501. to provide a setting for practicing computational skills; | 6 | 8 | 9 | 8.5 | 9 | 9 |
| 502. to develop methods of thinking and logical reasoning; | 1 | 1 | 1 | 1 | 1 | 1 |
| 503. to identify students who possess mathematical talent; | 10 | 10 | 10 | 10 | 10 | 10 |
| 504. to learn how to read mathematics; | 9 | 6 | 8 | 7 | 7.5 | 8 |
| 505. to apply recently taught mathematical ideas; | 5 | 5 | 6.5 | 4 | 5 | 6 |
| 506. to acquire skills necessary for living in today's world; | 2 | 2 | 3 | 8.5 | 6 | 5 |
| 507. to develop the skills to approach new topics in mathematics independently; | 7.5 | 9 | 6.5 | 6 | 4 | 4 |
| 508. to develop creative thought processes; | 4 | 4 | 4 | 2 | 2 | 2 |
| 509. to acquire problem-solving techniques that are vital to having a well-rounded education; | 3 | 3 | 2 | 5 | 3 | 2 |
| 510. to enhance the ability to apply mathematics in science. | 7.5 | 7 | 5 | 3 | 7.5 | 7 |

concepts, and, indeed, all the remaining components of the usual mathematics curriculum. The lack of precision in talking and thinking about the teaching and learning of problem solving was one of the primary reasons that problem solving was identified as one of the major curricular strands for which preference and priority information was sought.

Two ten-item clusters, one for secondary school content and the other for elementary school content, were designed. The teams of experts who helped write the items opted to identify heuristics and problem-solving strategies as the curricular content for problem solving. The questions inquired whether a given heuristic or technique should be taught to either elementary or secondary students. The graph in figure 1 indicates whether the technique *should definitely be included*(2), *should be included* (1), *undecided*(0), *should not be included*(−1), or *definitely should not be included* (−2). Color corresponds to mean responses relative to the elementary school curriculum, and black corresponds to secondary school curriculum. The following chart lists the items and their code letters as shown in the graph.

| *Code* | *Item* |
|---|---|
| J | Categorize problems into specific types (e.g., age, distance-rate-time), then teach a method of solution for each type. |
| K | Generate many possible answers using a calculator or computer, then check to see which one meets the conditions of the problem. |
| L | Write and solve a simpler problem; then extend the solution to the original problem. |

M    Explore the problem by using flowcharts.

N    Translate the problem into number sentences or equations.

P    Guess and test possible solutions.

Q    Start with an approximate answer and work backwards.

R    Draw a picture diagram or graph to represent the problem situation.

S    Construct a table and search for patterns.

T    Teach primarily global problem-solving ideas (e.g., read, plan, work, check).

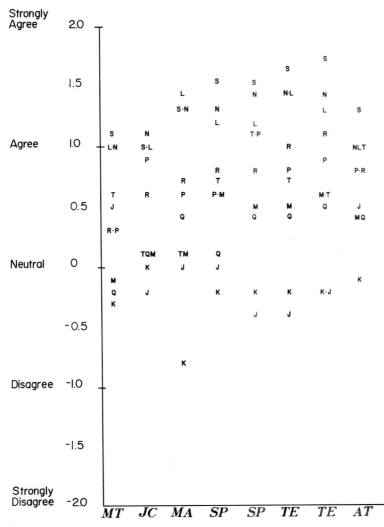

Fig. 1. Relative attractiveness of content alternatives in the problem-solving curriculum for selected populations. Color = elementary content; black = secondary content.

CONCLUSION 3. *The strategies of (1) translating a problem to an equation, (2) constructing a table and searching for patterns, (3) drawing pictures or diagrams to represent a problem, and (4) solving a simple problem first and extending the solution to the original problem were identified as appropriate content for both the elementary and secondary school levels.*

Interestingly, starting with an approximate answer and working backward was perceived more positively for elementary school learners than for students in the secondary school. Indeed, the MT sample tended to regard the strategy of working backward as inappropriate for curricular-related content. Items in other curricular strands and clusters indicated that this strategy was perceived as even more inappropriate if the use of the calculator or computer as a tool was included.

Categorizing problems by type and teaching a method of solution for each type was a content strategy viewed more positively by the MT and AT samples—primarily teachers—than by teacher educators, supervisors, or teachers of advanced mathematics. Why would teachers at the school level value specific strategies for given problem types more than the other populations surveyed? We conjecture that teachers at the elementary and secondary levels are more aware of the reality that "getting a right answer" is a primary motivation for many of their students. It should be noted that *all* populations felt it was more important to include global problem-solving ideas (e.g., read, plan, work, check) in the curricular program at every level than to employ the strategy of categorizing problems by types.

What resources are needed to teach problem solving? Responses show that respondents were positive about any resource that would facilitate teaching problem solving. Extremes of response were produced by items concerning printed materials. Those involving special problems for special populations—girls or ethnic minority students, for example—received low rankings. Low ranking was also given to text materials that included all the verbal problem-solving activities within a single chapter. Supplementary sources of problems, particularly those relating to real life or to applications, ranked high. Text modules directed toward establishing specific problem-solving heuristics ranked high also. The availability of computers for problem exploration and devices to model problem situations and solutions was also highly desired.

We do note for each of the curricular strands except computer literacy and problem solving that there was a cluster of items inquiring about appropriate use of the calculator. The sixty-five items were classified into six generic types—checking, nonroutine computation and problem solving, routine computation, developing ideas, homework, and tests. A substantial percentage of all populations perceived checking as the most appropriate use of the calculator. Using the calculator as a tool for problem solving and for nonroutine computation ranked second for most populations.

The professional samples responded to a cluster of items concerned with appropriate teaching methodology for problem solving. The item choices were described in terms of the methodology being a characteristic of the instructional materials. Respondents were to indicate whether the implicit methodology would influence them to buy or to use the materials. Methodologies of extreme types tended to be ranked low. For example, the item concerning long problems taking more than a single class session to solve (e.g., a USMES style of problem) was given low preference by the professional samples. Interestingly, these same samples were negative about methods that either required reading or de-emphasized reading.

*Only problems which students can answer quickly are assigned,* an extreme methodology, was strongly rejected by the PR, SB, and PT samples. Two other items reflecting traditional practices received high acceptance from these same three groups. One asked the respondent to consider this practice: *Short problem-solving sections are included after each mathematical topic is taught.* The other statement requiring reaction was this: *Students are shown how to solve a problem and then similar practice problems are assigned.*

Perhaps the most tenable conclusion concerning methods of teaching is—

CONCLUSION 4. *No extreme positions concerning teaching methodology for problem solving were preferred by any group.*

A few particular items of methodology deserve comment, however. All professional samples preferred assignments that "challenge students to think." Although "projects that involve real-life problem situations" was a methodology statement highly ranked by all professional samples, the older the students taught by a sample, the lower the ranking. That is, the methodology involving real-life situations was most important to the AT sample, less important to the MT and JC samples, and least important to the MA sample.

The PR, SB, and PT samples responded to the two items *Ideas or procedures are developed through real-life problem situations or activities* and *Each new mathematical topic is introduced with a problem to be solved.* Although support across samples was stronger for the former than for the latter (92.3 percent compared to 75.6 percent positive responses), both items were supported by a substantial majority of the three samples.

Likewise, all the professional samples reacted positively to item 532: *Problems are used to introduce mathematical topics.* An analogous item appeared in the methods cluster for each of four other curricular strands—algebra, measurement, ratio and proportion, and probability and statistics. The graph in figure 2 indicates the positive perceptions in each of the five mathematical settings. The preference data about methods indicated the following:

CONCLUSION 5. *Using a problem as the means or the vehicle to develop and introduce mathematical topics is a preferred methodology.*

For many of the preferences indicated by the data for other portions of the questionnaires, we have wondered to what extent the preferences are primarily an affirmation of beliefs about current practice. However, we know of little evidence that the indicated preference for introducing ideas through problems is typical of current classroom practice. Perhaps the appropriate question is, "If this is a widely held belief about the role of problem solving in teaching mathematics, then what situational variables in the school setting or in the mathematics classroom restrict using problems as vehicles for the development and introduction of mathematical concepts?"

One cluster of items was developed for each of the nine curricular strands about very specific issues. Most of the items concerned the type of student and the timing of instruction; thus, the cluster was designated by the label "Who/Time." The items concerned requirements, special courses for different categories of students, com-

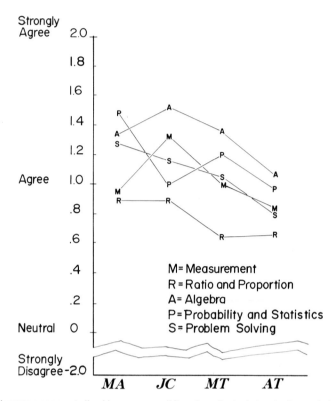

Fig. 2. Average responses to "problems as means" items in methods clusters for five curricular areas for the MA, JC, MT, and AT population samples

petencies that should be expected at different points in the curriculum, and other is-sues of a very specific nature.

The Who/Time cluster for problem solving yielded data indicating a great deal of ambivalence within populations. Many of the items generated extreme reactions. Four of the items produced average responses that were among the most negative for the entire 705 items in the preference survey item pool:

542. Problem solving is a function of intelligence and cannot be taught except to gifted students.

545. Problem solving is important only for college-bound students.

546. Different problem-solving courses should be offered for girls.

549. Problem solving should not be taught in the elementary grades.

When these responses are considered in conjunction with the responses to all other items about problem solving, the following conclusions are justified:

CONCLUSION 6. *Problem solving is important for all students and should begin early in their mathematical experience.*

CONCLUSION 7. *A modification of the mathematics curriculum to provide unique problem-solving experiences for special groups, such as women, the college bound, or ethnic minorities, is perceived as inappropriate.*

Respondents in the six professional samples exhibited considerable ambivalence about how to structure the curriculum to accomplish instruction in problem solving. Response patterns for two items are given below in percents (the five-choice response alternatives have been collapsed to three).

541. A separate problem-solving course, lasting at least one semester, should be required of all students before high school graduation.

| Population | Supportive | Neutral | Against |
|---|---|---|---|
| AT | 57.4 | 16.8 | 25.8 |
| MT | 42.1 | 15.1 | 42.8 |
| JC | 48.5 | 15.2 | 36.4 |
| MA | 30.0 | 13.3 | 56.7 |
| TE | 28.8 | 10.2 | 61.0 |
| SP | 24.0 | 12.7 | 63.4 |
| Total | 40.5 | 14.3 | 45.3 |

547. All problem solving should be done within existing mathematics courses.

| Population | Supportive | Neutral | Against |
|---|---|---|---|
| AT | 26.4 | 10.8 | 62.7 |
| MT | 37.9 | 22.7 | 39.5 |
| JC | 34.3 | 17.6 | 50.0 |
| MA | 30.1 | 23.3 | 46.6 |
| TE | 45.7 | 11.9 | 42.3 |
| SP | 59.3 | 4.2 | 46.5 |
| Total | 37.1 | 14.7 | 48.2 |

Of course, there are in each item a few "flag" words having nothing to do with problem solving that could account for the reactions to these items. Many individuals react strongly to the words *all, requirement,* and, in the case of item 547, *existing.* Nevertheless, there is a segment of each population favoring a special problem-solving course and a corresponding group that feels it best to embed problem solving within the framework of a broader course. Perhaps this is the type of curricular issue that is in need of a careful evaluative study to analyze the comparative payoffs and liabilities for the two curricular structures.

Should instruction on problem solving be interdisciplinary? Approximately 60 percent of the respondents to the Who/Time cluster favored the development of an interdisciplinary problem-solving course, notwithstanding the experience of curricular development projects such as USMES.

Finally, one Who/Time item that is perhaps most appropriately classified as a content cluster item concerned finding problems within situations. No qualifiers were offered for the word *situations;* individuals provided their own interpretations. A clear majority of each sample agreed that this was an important component of the problem-solving curriculum strand of a school.

The second-round survey—on priorities—required a respondent to assign priorities, one through five, for expending resources to five topical areas of mathematics education. The first set of five to feature problem solving listed for contrast the topics of measurement, fractions, decimals, and whole-number computation. The resource was money to be expended for curriculum development.

Problem solving was given highest priority across all the samples, whether professional or lay, with 67 percent of the total respondents ranking it highest. Only 3 percent of all respondents gave problem solving the lowest priority in the forced choice among the five topics. When queried concerning the reason for the high ranking given to problem solving, 56 percent of those who ranked it highest avowed that "it is absolutely crucial that all students develop skills in this area."

Time was the second resource featured in the priorities items concerned with problem solving. The priority assignment of five instructional ideas (solving word problems, drill and practice on basic number skills, exploring enrichment topics, studying applications of mathematics, and building an intutitive base for algebra and geometry) was in terms of how best to use an additional fifteen minutes a day on instruction in mathematics at the elementary school level. Additional time for word problems was the most chosen of this forced selection of priorities for the AT, SP, and TE samples and second for the remaining samples—MT, PR, SB, and PT.

The data for the problem-solving items on the second-round surveys are consistent with the findings for the preference surveys. The second-round survey provides more direct evidence of the comparative ranking of problem solving in contrast to other curricular areas than is possible from the inferred rankings of the preference survey. The priorities survey provides strong supportive evidence for the wisdom of featuring problem solving as the central core of NCTM's curriculum recommendations for the 1980s. Frequently, curriculum recommendations in the past have not led to constructive modifications of school mathematics programs. Often the reason that a set of curricular recommendations fails to promote constructive changes is a lack of support for the ideas. It is to be hoped that the PRISM survey data indicate a sufficiently powerful commitment to problem solving to accomplish needed changes in the curriculum as we begin the decade of the 1980s.

# 7

# *Pictorial Languages in Problem Solving*

## Joel Schneider
## Kevin W. Saunders

$A$T ANY level—elementary school, high school, or college—a major difficulty in teaching problem solving lies in convincing students to record the details of a problem on paper. Those who consistently record pertinent information may be more likely to become successful problem solvers.

Some of the unwillingness of students to commit details to paper may stem from a resistance to the typcial requirement that information presented in a word problem must be translated into numerical expressions. The symbols and grammar of mathematics constitute an unfamiliar language, and students differ in the speed and facility with which they attain an understanding of it. When the language of number expressions is unfamiliar, students must labor to express their understanding of a problem in a foreign language, as it were, rather than being free to concentrate on the problem itself.

An alternative approach in the early teaching of problem solving is to provide a pictorial language with which children can record information. In our experience, such a language encourages them to put information on paper. Moreover, they tend to record information in ways that *they* find useful. Then, having instilled good habits of information processing at the primary level, we can offer increasingly sophisticated problems and introduce symbolic language as students develop.

We shall describe two examples of students using pictorial language in problem solving. In the first example, the problems were part of a first-grade curriculum (Comprehensive School Mathematics Program 1978a). Pupils used dots, arrows, and strings to display numerical relations extracted from problem statements. In the second example, whose problems were part of a fifth-grade curriculum (Comprehensive School Mathematics Program 1978b), the language was enriched by giving the dots and arrows geometric interpretations. Both classes were representative of their school district.

## Pictorial language in the first grade

Early in this first-grade class, pupils became familiar with the use of dots, strings, and arrows in telling stories. For example, the teacher drew strings of two different colors, each enclosing some dots. The dots in the left string represented apples and those in the right string, horses. After asking students to count the apples and the horses, the teacher drew colored arrows to show which apples are fed to which horses (fig. 1). Discussion centered on which horse eats the most apples, how many horses eat no apples, and so on.

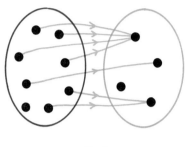

Fig. 1

In other lessons the students invented stories to fit pictures such as the one in figure 2. The following interpretations were typical of their responses: children in the string and pets outside, with arrows for "you are my pet"; dogs inside and bones outside with arrows for "chews." Throughout the year the children experienced many problems of this type.

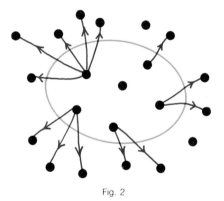

Fig. 2

During the second semester, they were given a series of three word problems at two-week intervals. The teacher read each problem aloud twice, and then the pupils worked individually with paper and colored pencils to draw whatever they thought would help solve the problem.

1. A friend of mine owns a gas station. This morning 7 cars came in and each car needed 4 new tires. How many tires did my friend sell?

2. Six octopi walked into a shoe store and they all needed new shoes. How many shoes were needed altogether? (Octopi had been discussed previously.)
3. I have 24 bottles of soda and I want to put them in packs that hold 8. How many packs will I need?

These problems may seem advanced for the first grade, but with the use of dots, strings, and arrows the children were successful in solving them. Almost everyone solved the first problem, and between two-thirds and three-fourths of the class solved the other two. Several of those who failed to find the correct answer for the second problem gave such answers as 46 and 49, indicating a counting mistake. Figures 3–5 are reproductions of drawings they used to solve the problems. The solutions exemplify three strategies using dots, strings, and arrows to solve such numerical problems.

To solve problem 1, Chris drew seven large dots for the cars and four dots for each car's tires. Arrows assigned the tires to the cars (fig. 3). To solve problem 2,

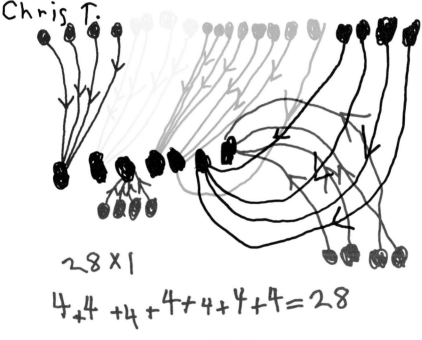

Fig. 3

Meredith drew a string for each octopus and placed eight dots—the shoes—in each string. Then she noticed that she had drawn too many strings, and so she crossed two of them off (fig. 4). For problem 3, Shannon drew dots for the soda bottles and strings for the eight-packs (fig. 5).

The pupils turned each of these problems into a counting situation. As they finished, they whispered their answers to the teacher, and those who finished early were asked to write number stories. Most first-grade students would not know the

Meredith

Fig.4

Fig. 5

Shannon

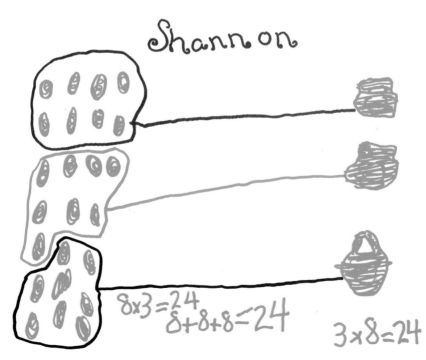

$8 \times 3 = 24$
$8 + 8 + 8 = 24$
$3 \times 8 = 24$

multiplication or division facts involved in the problem, much less be able to write number stories, without the support of their drawings. However, the experiences did show that first graders are able to use pictorial language to express the details of a problem. Having recorded the details, they can concentrate on the counting or arithmetic involved.

### Pictorial language in the fifth grade

Geometry often provides a helpful setting for solving arithmetic problems. A group of fifth-grade students were exploring parallelism. Provided with a simple rolling straightedge, called a translator (see fig. 6), they learned to constuct parallelograms

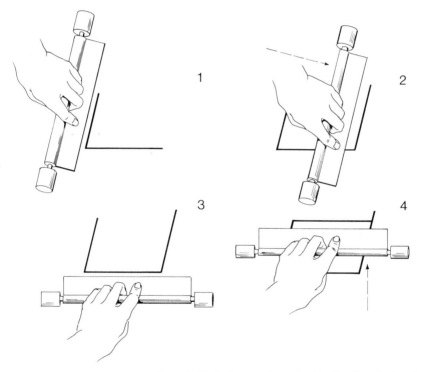

Fig. 6. Aligning the translator with a line and rolling it allows one to construct families of parallel lines. In particular, one can easily construct a parallelogram, as illustrated. Several models of translators are commercially available.

and to perform a variety of constructions based on parallelograms, some of which are shown in figure 7.

The class learned of a fruit market so small that it offered only two items, apples and blueberries. The small size of the market encouraged a friendliness between the proprietor and customers that was expressed in a policy of allowing returns and exchanges of apples and blueberries. Initially the students developed a technique of recording transactions on a lattice. In figure 8, a red arrow represents the purchase of one kilogram of apples and a gray arrow the purchase of one kilogram of blue-

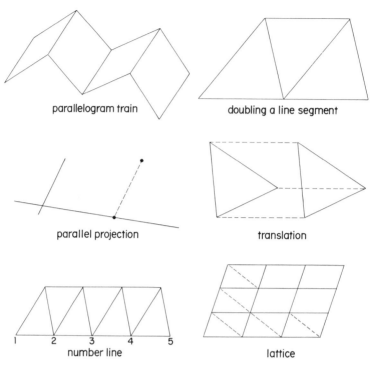

Fig. 7

berries. Each dot represents the result of a specific purchase: "$1a$" for one kilogram of apples, "$1b$" for one kilogram of blueberries, "$3a + 2b$" for three kilograms of apples and two of blueberries, and so on. Of course, many purchases are possible for which the dots are not shown: $0.5a$ and $3.25a + 1.50b$, for example.

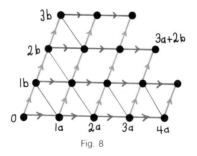

Fig. 8

As the students became more familiar with the device, they were able to use it to represent a full range of transactions. Paula constructed an apple axis and a blueberry axis (fig. 9). Dots representing one kilogram of apples and one kilogram of blueberries determined the scale of the lattice. Using a translator, Paula constructed parallelograms to locate dots to represent several additional transactions. (*Note:* "$\hat{1}$" is a notation used interchangeably with "$-1$" for "negative one.")

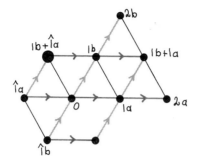

Price calculations were the next complication. Given unit prices for apples and blueberries, the class calculated the costs of several transactions. The lattice was a direct support for the mental arithmetic involved. For example, to find the price of two kilograms of apples and three kilograms of blueberries, one can trace any path from the dot for 0 to the dot for $2a + 3b$, adding the cost of each kilogram in turn. To find the cost of noninteger quantities, the students relied on their familiarity with graduations of number lines.

Consider these two problems:

**Problem:** With apples at 30 cents for one kilogram and blueberries at 40 cents for one kilogram, what can be bought for exactly one dollar?

**Problem:** Find solutions for the equation

$$30x + 40y = 100.$$

Of course, the problems are essentially the same. But the first is likely to be more interesting to students because of the market context. To study the first problem, Jenny constructed a lattice and marked points in red representing purchases of one dollar (fig. 10).

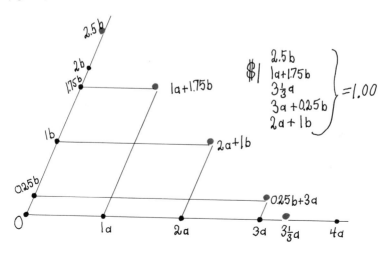

After several solution points were located, some students guessed that all solutions lay on the same straight line. In testing their conjecture, they calculated the cost of transactions·represented by other points on the line. Choosing a purchase point, they projected it parallel to the blueberry axis onto the apple axis and vice versa (fig. 11). The projection onto the apple axis showed the amount of apples in the purchase; the projection onto the blueberry axis showed the amount of blueberries. With this information they checked the cost of the purchase. Of course, the conjecture was sound, and many solutions to the problem were generated from the lattice.

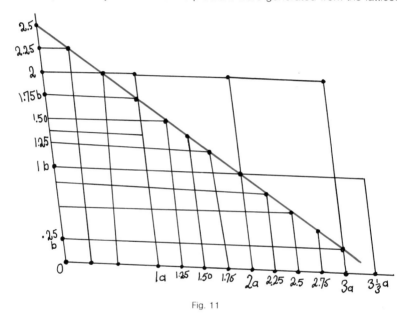

Fig. 11

To elaborate the problem, students considered transactions at 60 cents and transactions involving no money—even exchanges. For example, with prices at 20 cents for one kilogram of apples and 30 cents for one kilogram of blueberries, three kilograms of apples could be traded for two kilograms of blueberries. And we can represent the trade by $\hat{3}a + 2b$. That is, the net result of the trade in terms of money is 60¢ + 60¢ = 0¢.

After locating the points for many of these transactions, Vonda distinguished two lines on her lattice as shown in figure 12. That the two lines were parallel was quickly conjectured by the students and verified experimentally with the translator. Of course the lines cannot cross, for then there would be a transaction with two prices.

In constructing the lattices, students enjoyed the freedom of determining the layout for themselves. At the same time, they were able to construct drawings with satisfying accuracy by using the translator. Their careful construction of a geometric model of the market situation resulted in a deeper level of involvement than prepared lattices or purely arithmetic approaches would afford. In fact, the geometry interacted with the arithmetic by displaying solutions to the problems. Moreover, the display itself suggested additional solutions.

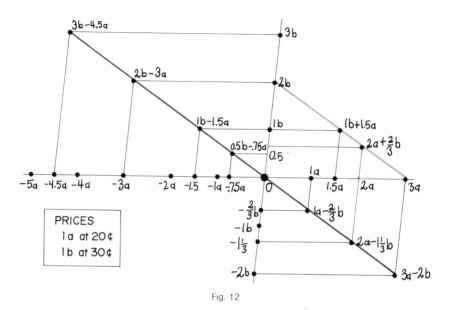

Fig. 12

In both situations, pictorial language engaged students in problem solving and provided a means for a fruitful expression of information. Many of the first graders were able to move beyond the pictorial representations to numerical expressions. The fifth graders were led by the geometry to observations and conclusions not explicit in the statements to the problems. Whether as a bridge to a numerical language or as a rich context for exploring a problem, pictorial languages prove to be useful, rewarding tools in solving problems.

### REFERENCES

Comprehensive School Mathematics Program. *Mathematics for the First Grade.* St. Louis: CEMREL, 1978*a.*

———. *Mathematics for the Intermediate Grades, Part IV.* St. Louis: CEMREL, 1978*b*.

*8*

# Polya Visits the Classroom

## Linda J. DeGuire

GEORGE POLYA is a master problem solver and teacher of problem solvers. Even though his style of teaching cannot be captured in a single article, perhaps not in any written communication, this essay is an attempt to present something of that style.

First, Polya teaches by example. He is a fellow problem solver in his teaching. He does not present problems as solved but *to be* solved. Working at the chalkboard, he elicits a solution from a class by using leading questions and suggesting productive strategies. At times Polya steps out of the role of fellow problem solver to look back over the part of the solution just completed, becoming an outside commentator on the strategies and processes being used. As commentator, he physically steps a few feet to the side and discusses what is taking place. These commentaries illustrate an important distinction between merely solving a problem with a class and teaching problem solving; they emphasize the methods being used rather than a particular solution of a particular problem.

Second, he gently weaves heuristic suggestions into each problem-solving episode and into an entire course. He does not give a series of problems, each requiring the student to use the same heuristic suggestion. Rather, he begins with a few heuristics and, as the students assimilate these, gradually introduces more.

Finally, Polya assumes the role of commentator not only within the problem-solving episode but also at its end. He constantly exposes his students to reflective problem solving; he reflects on the problem and its solution, looking for other methods of solution, generalizing the results and strategies, and generating new problems.

The flavor of Polya's teaching can perhaps best be conveyed as he would teach—by example. The two episodes that follow are attempts to illustrate his style by imagining how a teacher might use two problems, following Polya's technique, to teach problem solving to middle school students. One problem is quite simple, and the episode contains only a few heuristic suggestions. The other problem is more complicated and contains more suggestions. In the episodes, the roles of commentator and teacher have been separated to emphasize their distinction; however, if Polya were teaching, he would actually play both roles.

70

# Episode 1

*The setting is a fifth-grade classroom. In previous problem-solving episodes, the teacher has introduced the students to such ideas as the conditions and the unknown of the problem. In this episode, the teacher is emphasizing the four-phase plan of Polya's* How to Solve It *and, especially, the use of a figure.*

*Teacher (T):* Last night I finished making my guest list for the dinner party I am going to give next month. Because there will be thirty people, I will need to borrow some card tables, the size that seat one person on each side. And I wish to arrange them in a long row, end to end. Of course, I want to borrow as few tables as possible. How many card tables will I need? Try solving my problem.

*[Long pause. A few minutes for students to tackle the problem.]*

Notice that the teacher has stated the problem in a small story. Polya frequently uses stories to introduce his problems. Stories are a means of getting students involved in the problem.

Also notice the long pause, well placed to entice students to become involved. The pause leaves the situation open ended to allow the students to try or suggest whatever strategies or questions they might naturally use. It also allows the teacher a few minutes to see what the students are doing.

*T:* Let's see what we are coming up with.

*Commentator (C):* Remember that the first phase in solving a problem is understanding the problem.

*Mike:* You can forget all of that because it's such a simple problem. You can put four people at each table. So, you just divide four into thirty to get 7½—you will need 7½ tables.

*Pete:* Mike, you can't put four at each table!

*Mike:* Why not?

*Pete:* Because she said she wanted to put the tables in a row, end to end. Besides, who ever heard of half a table!

*T:* Yes, Mike, you seem to have forgotten some of the information in the problem.

*C:* Mike is reminding us that we have not stopped to be sure we understand each condition of the problem. What are we asked to find? That is, what is the unknown? It is the number of card tables needed. What information is in the problem?

*Mike:* Thirty people at the party. And you are putting the tables end to end. So, I guess you can't put four people at each table.

*C:* True, Mike. But what about half-tables?

*Pete:* Well, if you come up with a fraction, like 7½ tables, you really need eight tables.

*C:* In other words, you will use only counting numbers as an answer. What we have been doing is being sure we understand the problem—that what we want to find (the unknown) is the number of tables needed, that what we have are thirty people to be seated, and that the conditions are that the tables will be put end to end and that you can use only a whole number of tables. Let us turn now to finding a plan for solving the problem.

The commentator is reminding the students of the four-phase plan for solving problems. This dual role of the teacher—as fellow problem solver and as outside commentator on the processes—is at the heart of Polya's teaching style.

*T:* How could we go about solving this problem?

*Jan:* I tried drawing little squares to represent the card tables.

*T:* Excellent idea, Jan! How about coming up and being our artist for this picture?

*C:* Drawing a figure is a very useful idea in solving many problems. It will usually help us to visualize our information and test possible solutions.

*[Jan draws a figure like that shown in fig. 1.]*

Fig. 1

*Jan:* But I don't know how long to make the row yet.

*T:* Well, how can we represent the number of tables?

*Lynn:* We could use a letter.

*T:* What letter would be a good choice, Lynn?

*Lynn:* We could use *t* for "tables."

*Jan:* Or *n* for the "number of tables."

*C:* Either letter is a good choice to stand for our unknown because either letter reminds us of what it represents.

*T:* Now, Jan, can you use our unknown to finish drawing the figure? *[Jan adds to fig. 1 to make fig. 2]*

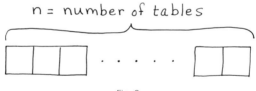

Fig. 2

*C:* By putting our unknown onto the figure, we are making the figure express more parts of the problem. Has the figure suggested a plan of solution to anyone yet?

*Terry:* Yes. You can just count how many people are on each side. So, you will need fifteen tables.

Terry has given an incorrect solution to the problem. The teacher now wants to *lead* the students to realize that the number of tables can still be reduced.

*T:* Terry, show us on the figure how you count the people. *[Terry draws fig. 3 and counts aloud as she numbers the places at the table.]*

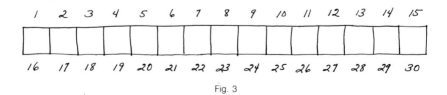

Fig. 3

*C:* Ah! Apparently we have solved the problem. Let's go back to be sure that our solution is correct and to see if we could have solved it any other way.

*T:* So, you say I will need fifteen tables. Is this the *least* number I could use?

*Lynn:* There's no one sitting at the ends. We could put two people there. But then we wouldn't need so many tables.

*T:* How many would we need?

*Mike:* We would need only fourteen tables because we can take one table away and put those two people at the ends, like this. *[He triumphantly erases one table from fig. 3 and re-numbers it to look like fig. 4.]*

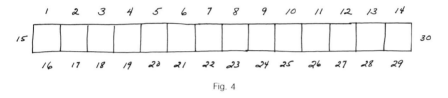

Fig. 4

*T:* It looks like checking our first solution was a good idea. We hadn't solved the problem correctly the first time. But are we *sure* we have the minimum number of tables now?

*Terry:* Yes, because there aren't any more unused places.

*C:* So, we have checked and corrected our solution to the problem. But surely there is nothing special about having thirty people!

*T:* What if I had invited 42 people? Or 100 people? Or 500 people? We'd certainly have a lot of little squares to draw!

The teacher is beginning to use the problem to generate a new problem, specifically, a generalization of the original problem. Polya rarely leaves a problem without generalizing it. Such generalizations, of course, would have to be adapted to the level of the students. Thus, in this example aimed at middle school children, the generalization is stated as a verbal rule rather than as a variable expression.

*Pete:* Let's just do what we did before—draw it wrong and then erase a table to correct it.

*T:* Can you say that in some kind of a rule or process?

*Pete:* Ummm . . . well, first, you take the number of people and divide by 2. Then just erase a table.

*Lynn:* Erasing a table would be just like subtracting one.

*T:* So, let's write our rule under the first figure, the one that shows the unknown: "Divide the number of people by 2 and then subtract 1." Let's check it for thirty people: $30 \div 2 = 15$ and $15 - 1 = 14$ tables.

*Jan:* I was going to write the rule differently.

*T:* Let's hear it, Jan! It may work also.

*Jan:* Well, the two end tables have three people each. The rest of the tables have two people each. So, take the number of people, subtract six, and divide by two. But then we have to add on the two end tables.

*T: [Genuinely surprised.]* That's very good, Jan! I hadn't thought of doing it that way. Let's check it with our case of thirty people: $30 - 6 = 24$; then $24 \div 2 = 12$ tables that seat two each; then these 12 tables + 2 end tables = 14 tables. It works! *[Writes rule on board and labels it "Jan's rule."]*

Frequently, students will solve a problem in a way new to you. Such an experience helps to widen your own stock of strategies and solutions and so needs to be remembered somehow. Polya sometimes suggests keeping a diary of one's problem-solving experiences. In this context, such a diary could translate into a problem file of 3″ × 5″ or 5″ × 8″ cards with teaching notes on the front and different solutions on the back. The teacher could add Jan's solution to the back of her card.

*C:* So, we have written two rules that seem to work for any number of people. Let's check and see if they work for other numbers.

*[The teacher and class check both rules for several numbers, such as 42, 8, and 4.]*

At this point, Polya would not let the problem drop but would continue using the problem to discuss the strategies and methods used and to generate related problems by varying the conditions. Here are a few problems he might generate from this first example: Could I arrange the tables in a U or an L or some similar (but still connected) shape and thereby reduce the number of tables? If I decided to make two rows of tables instead of one long row, how many tables would I need? What if I made three rows? What is the minimum number of tables I would need regardless of how I arranged them? What is the minimum number of tables I would need if I wanted no unused sides left over? If I had an odd number of people, how might the original solution change? What would happen to the answers of these related problems if I had an odd number of people?

## Episode 2

*The setting is an eighth-grade classroom. Notice that the teacher brings a wide variety of heuristic suggestions into this problem-solving episode, since these students are more mature problem solvers than those in episode 1. Particular emphasis is given to two suggestions: (1) specialize the problem and (2) solve the problem another way.*

*Teacher (T):* Recently we have been solving some problems that involve number patterns. Look at this pattern. *[Draws fig. 5.]* Do you recognize it?

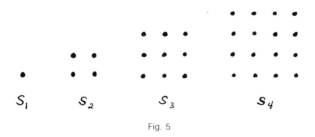

$$S_1 \qquad S_2 \qquad S_3 \qquad S_4$$

Fig. 5

*Jane:* Those are the square numbers.

*T:* That's right. This is a geometric way of representing the sequence of square numbers. So $s_n = n^2$. But what is so special about the geometric figure of a square? What if we represented a sequence of numbers by triangular arrays of dots? Let's try it and see what we get. Here are the first three triangular numbers. *[Draws first three arrays in fig. 6.]* What do you think the fourth one will look like?

*John:* Put a fourth row that has four dots.

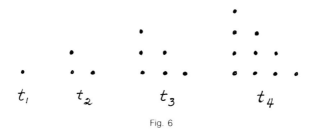

Fig. 6

*T:* So, the fourth one looks like this. *[Draws fourth array in fig. 6.]* Good.

*Commentator (C):* Remember that suitable notation can help in solving a problem.

*T:* How might we label these arrays for easy reference?

*Ted:* Let's use *t* to stand for "triangular number." Then we can use a subscript to number the terms in the sequence.

*T:* Good, Ted. So I'll put the symbol below each array we have drawn on the board. *[Puts labels under arrays in fig. 6.]* If I draw an array for $t_8$, what would it look like? . . . Jane?

*Jane:* It would have eight rows of dots.

*T:* And how many dots in the last row?

*Jane:* Eight.

*C:* It is important that we understand our symbol. Our symbol tells us not only how many rows of dots are in the array but also how many dots are in the last row.

*T:* What do you think my next question will be?

Problem finding is an important predecessor to problem solving. Since the students have already solved similar problems, Polya would expect them to recognize the similarity and, by analogy, to ask similar questions.

*Cindy:* Can we find the pattern? So, what is $t_{100}$? Or $t_{2000}$? Or $t_n$?

*T:* Good questions, Cindy. Does anyone have an idea how big $t_{100}$ is? *[Pause.]* No guesses? . . . Bill?

Polya strongly encourages students to guess. Guessing can involve the students by getting each one to defend his or her own guess. A guess can be used to find bounds for the solution and can help students understand the conditions of the problem. Guessing or predicting the next term in a sequence is a way of testing a generalization without verbalizing it.

*Bill:* $t_{100}$ has to be less than 10 000.

*T:* Why is that, Bill?

*Bill:* Because 10 000 is 100 × 100. That's how many dots there would be in a square array with 100 on each side. A triangular array has fewer dots than a square array.

*T:* Good idea, Bill.

*Bill:* In fact, the triangle looks like exactly half of the square with the same number of rows. So $t_{100}$ ought to be 50 × 100, or 5000.

*Sue:* That can't be right. Because then $t_4$ would be 2 × 4, or 8. But $t_4$ is 10. And that means that $t_{100}$ ought to be more than 5000. So the pattern must be more complicated.

*T:* Let's see if we can see a pattern. We need to look at some cases.

*C:* Frequently, the use of special cases in a systematic way can lead to a generalization.

*Kay: [Going to the chalkboard.]* Let's make a table of *t* numbers. *[Starts putting fig. 7 on the board and the rest of the class helps her put in additional entries.]* Once you get started, it's easy to keep going without drawing any triangular arrays at all.

*T:* Why is that?

*Kay:* I'll draw arrows on the table to show you. *[Draws arrows as in fig. 8.]* 1 + 2 = 3. Then 3 + 3 = 6. Then 6 + 4 = 10. Then *[several students joining her, in unison]* 10 + 5 = 15. And so on.

Fig. 7                Fig. 8

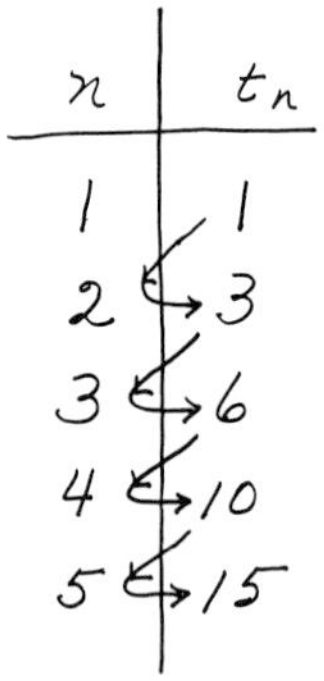

*T:* So, to get $t_5$, you add 5 to $t_4$. Is that right? *[Nods of agreement.]* What will $t_6$ be?

*John:* $t_6$ will be $t_5$ + 6. That is, $t_6$ is 15 + 6, or 21.

*T:* How can you be sure $t_6$ is 21?

*John:* Well, I could draw a picture. *[Draws $t_6$ and counts the dots.]*

*T:* Good. Now, what about $t_{100}$?

*Kay:* It will be $t_{99}$ + 100.

*T:* Right. And $t_{2000}$?

*John:* $t_{1999}$ + 2000.

*Cindy:* So, to finish off the problem, $t_n$ will be $t_{n-1} + n$.

*C:* Let's look back over our solution. Our plan was to write the information in a table and look for a pattern. That we have done. We have concluded that $t_n = t_{n-1} + n$. But is this pattern a good way to find $t_{2000}$?

*Ted:* No. We need to know $t_{1999}$ and we don't know it. We'd have to know all the previous *t* numbers to find $t_{2000}$.

*C:* Remember that thinking of a similar problem can frequently suggest a plan for solving a problem. Can you relate this *t* sequence to any other sequence we have studied?

*John:* We've seen that sequence someplace before but I can't remember where.

*T:* Think about the pattern we have already noticed—each $t_n$ is the sum of $t_{n-1}$ and *n*. No one recognizes the pattern yet? *[While talking, the teacher adds the third column to the table in fig. 8 to make fig. 9.]* Now, look—$t_1$ is 1. Then $t_2$ is 3, which is 1 + 2. Then $t_3$ is 6, which is 1 + 2 + 3. *[Several students join in.]* So $t_4$ is 10, which is 1 + 2 + 3 + 4. And $t_5$ is 15, which is 1 + 2 + 3 + 4 + 5.

| $n$ | $t_n$ |
|---|---|
| 1 | $1 = 1$ |
| 2 | $3 = 1 + 2$ |
| 3 | $6 = 1 + 2 + 3$ |
| 4 | $10 = 1 + 2 + 3 + 4$ |
| 5 | $15 = 1 + 2 + 3 + 4 + 5$ |

Fig. 9

*T:* Good. Did everyone understand what we were doing? *[Some indicate confusion; so she explains the process again.]*

*T:* Now, that is the pattern for $t_5$. What is the pattern for $t_{100}$?

*Cindy:* $t_{100}$ would be $1 + 2 + 3 + \cdots + 100$. And I remember the shortcut for doing that! It's the way Gauss did it in that story you told us. You write the numbers twice, each time in a row, once increasing and once decreasing, like this. *[Draws fig. 10.]* Then you add the rows and get 100 of the 101's; so that's $100 \times 101$. But you've got each number twice. So you divide by 2. So you get $\dfrac{100 \times 101}{2}$. Let's see . . . umm . . . that would be 5050. So $t_{100} = 5050$.

$$\frac{\begin{array}{c} 1 + 2 + 3 + \ldots + 98 + 99 + 100 \\ 100 + 99 + 98 + \ldots + 3 + 2 + 1 \end{array}}{101 + 101 + 101 + \ldots + 101 + 101 + 101}$$

Fig. 10

*T:* Very good explanation, Cindy! You remembered that problem and how to solve it very well. Is everyone convinced that Cindy's method works? *[Some students hesitate.]* Well, it does agree with Bill's statement that $t_{100}$ is less than 10 000 and Sue's statement that it is more than 5000. What does Cindy's method give us for $t_n$?

*Sue:* Well, you would have two rows with the numbers from 1 to $n$ in them, and when you took the sum, it would be a row of $(n + 1)$'s.

*T:* How many would there be?

*Sue:* There would be $n$ of them. So the formula would be $t_n = \dfrac{n(n + 1)}{2}$.

*T:* Good. Let's show that both Cindy's method and the formula work for $t_5$. Remember we showed that $t_5$ is $1 + 2 + 3 + 4 + 5$. We already know that $t_5$ is 15 by drawing the triangular array and counting the dots. Also, it's easy to add up $1 + 2 + 3 + 4 + 5$ to show that the sum is 15. By Cindy's method we get this. *[Puts fig. 11 on board.]* So $5 \times 6 = 30$. But this is twice as much as we need. So $30 \div 2 = 15$. Putting these together, we have $\dfrac{5 \times 6}{2} = 15$, which is what the formula gives us. Now, Ted, if you used the formula to find $t_{100}$, what would it give you?

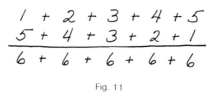

Fig. 11

*Ted:* $t_{100}$ is $\dfrac{100 \times 101}{2}$, which is 5050.

*T:* Good. That agrees with our answer when we used Cindy's method. And what is $t_{2000}$?

*John:* $t_{2000}$ is $\dfrac{2000 \times 2001}{2}$, which is 2 001 000.

*T:* Whew! That's quite large.

*C:* Looking back over our two solutions, we see that the first solution gave us a recursion formula $t_n = t_{n-1} + n$. This pattern was easy to see, but the formula was not very efficient. Our second solution involved seeing that this problem is just another form of one we had already solved: the problem of finding the sum of the first $n$ positive integers. However, there is another solution that might be more convincing to some of you. Bill, your idea of taking half a square gave an answer that was close, but not quite right. Can you see how to make it right?

*Bill:* Not very well. I don't see how the triangular array fits into the corresponding square array.

*T:* Look at the arrays of square numbers and of triangular numbers [*points to figs. 5 and 6*]. Let's use $t_4$ and $s_4$. How can we see $t_4$ in $s_4$? . . . No ideas?

*C:* Perhaps you could separate the figure by drawing an auxiliary line.

*Sam:* If you draw a diagonal in $s_4$, you get two triangular arrays plus the diagonal. [*Draws fig. 12.*] But the triangular arrays are smaller. I mean, you get $t_3$ on each side of the diagonal, not $t_4$.

*Jane:* I see it now—$s_4$ is two $t_4$'s put together . . . no, it isn't. We don't have enough dots in the square for that.

*T:* But, Jane, what happens if you put two $t_4$'s together?

*Jane:* You *almost* get $s_4$, but you have an overlapping part on the diagonal. I mean, you need the diagonal with each of the $t_4$'s.

*T:* Then, let's put an extra diagonal in the figure. It would look like this. [*Draws fig. 13.*] I'll draw a line where the two $t_4$'s come together. Now what do we have?

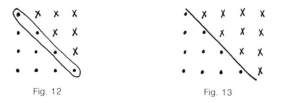

Fig. 12                              Fig. 13

*Bill:* It's a rectangle—five units long and four wide.

*T:* Good. So, how many dots does the rectangle have?

*Bill:* Twenty. And $t_4$ is half of it. So $t_4$ is 10.

*T:* And that checks with what we had before. Now, what would happen if you put two $t_n$'s together?

*Sam:* You'd get a rectangle that is $(n + 1)$ dots long and $n$ dots wide. Oh, I see! It would have $n(n + 1)$ dots. But $t_n$ is only half of that. So $t_n = \dfrac{n(n + 1)}{2}$, which is the same formula we had before.

*T:* Right. So now we have a geometric solution to the problem.

*Cindy:* That's a neat way! I like that one.

*T:* Yes. I find it especially convincing because it can be seen visually.

*C:* Let's look back over this last solution. Bill had a bright idea for a geometric solution, but it wasn't quite right. However, we were able to modify his idea by drawing in the diagonal of the square and by adding more dots to make the square a rectangle with two $t_n$'s in it. Then $t_n$ was half the rectangle. So, now we have three ways to solve this problem. Now certainly there is nothing special about triangular and square arrays and the number sequences they represent. We could consider pentagonal or hexagonal arrays or any polygonal shape. However, let's leave these for you to explore on your own or some other time together.

## Conclusion

These episodes are only glimpses into Polya's teaching style; in many ways they are contrived and artificial. For example, because of space limitations, the episodes do not include the many blind alleys that students attempt to follow, nor do they include Polya's patient guidance of students out of those blind alleys. Also, the two episodes do not illustrate his long-term intertwining of suggestions and strategies. However, the episodes do reflect his dual roles as fellow problem solver and commentator. Part of Polya's effectiveness as a teacher of problem solving rests on his skill in maintaining the proper balance between these roles. As seen in the episodes, the commentator asks questions that the students eventually will ask themselves. Of course, such independent problem solving is Polya's goal.

The development of problem-solving skill is a long-term goal. It takes a commitment to expose students to problem solving as often as possible. For both problem solving and the teaching of problem solving, Polya's advice is most appropriate—practice, practice, practice. Perhaps these glimpses into the master's teaching style will encourage other teachers as they develop an effective style in teaching problem solving.

### REFERENCES

Polya, George. *How to Solve It.* 2d ed. Princeton, N.J.: Princeton University Press, 1957.

———. *Mathematical Discovery.* 2 vols. New York: John Wiley & Sons, 1962, 1965.

# *9*

# *Improving Story-Problem Solving in Elementary School Mathematics*

## Edward J. Davis
## William D. McKillip

TEACHING children to solve "story" problems has been a difficult area in elementary school mathematics. These problems are important because they are the major vehicle—sometimes the only one—through which we address applications of mathematics. Because they have been, and continue to be, difficult, some students and teachers have developed negative attitudes toward them. Some teachers have never found a straightforward means of overcoming these negative attitudes and improving students' success in solving story problems; that is what we attempt to provide here.

The ability to solve problems is one of the most important objectives in the study of mathematics. Perhaps it *is* the most important objective, since the solution of problems—in mathematics, science, business, and daily life—is the ultimate aim of the study of mathematics. Problem solving has two different facets: one is understanding the problem thoroughly and selecting and applying mathematical notions that might lead to a solution; the other is getting the right answer. Although getting the right answer is highly desirable, it should always result from a careful analysis of the problem. An emphasis on getting the correct answer at any cost can lead to dubious approaches. Unfortunately, correct answers seem to be an all-important goal in the eyes of many parents and educators. Perhaps this is why children come to value them so highly. Let us consider some of the practices we employ, perhaps unknowingly, that de-emphasize the necessity of analyzing the problem.

One is the use of problem sets in which analysis assumes little importance—for example, sets in which all the problems are solved using the same operation, usually the operation the students are studying at the time. Students soon realize that a careful analysis of the problem is a waste of time, and they simply pick the numbers out of the problem and perform the operation. Somewhat better than these one-operation problem sets are those sets requiring more than one operation. Even in these sets, in which it is at least necessary to understand the problem and select the operation, it is assumed that solutions are reached by selecting and performing one or two arithmetical operations. Little opportunity is presented for developing strategies of searching, exploring, or trial and error.

Problem-solving lessons are still presented in which children are taught to look for individual words or phrases as the key to selecting the proper operation: "If the problem asks how many are *left*, you subtract." This is certainly a travesty of problem solving. The written problem, instead of being viewed as a description of a situation requiring mathematical analysis, is viewed as a set of words, only one of which is important in deciding what to do to obtain the solution.

Of course answers are important. But the best way to get correct answers is to analyze the problem carefully before plunging in. Suggestions for encouraging children to analyze problems will be discussed later.

## Improving Students' and Teachers' Attitudes

Many teachers do not feel very successful in teaching story problems; many students find story problems one of the more difficult challenges in mathematics and do not like them. Success leads to positive attitudes, and so we must begin with success. Throw out any consideration of grade level and start with very easy problems. Display or write simple problems that all the children in your group can solve.

Since long sets can be distasteful, give short sets—very short. Two, three, or perhaps four problems are plenty! However, give these sets of exercises frequently—three or four times each week.

The treatment for changing students' attitudes from negative to positive is to provide very frequent short sets of problems on which the students experience absolute success.

And what about you, the teacher? Who will help you with your attitude toward story problems? Don't worry about it. As you see your students succeeding in solving story problems and perhaps even liking them, your own attitude will improve; you will find it interesting and exciting to teach problem solving.

## Procedures for Improving Students' Skills

### Creating appropriate problems

Certain elements in the problem itself can influence the success or failure of a student in solving it. Problems can be manipulated in order to yield a high level of student success. We wish to create and use problems that are not trivially easy or beneath and dignity of the students. We wish to produce problems that combine challenge with a high rate of success and to manipulate problem elements to the students' advantage. Using problem 1 below, we shall present suggestions for doing this.

### Substituting small numbers

It is interesting to watch students when they face problems such as the following:

1. A sporting goods store has 247 baseballs worth $2.37 each and 142 softballs worth $3.84 each. What is the total value of the baseballs and softballs on hand?

Some students will look and look at such a problem without attempting a solution, even students you believe should be capable of solving it. When someone appears to be stuck, try presenting the same problem with very small, simple numbers:

2. A sporting goods store has 3 baseballs worth $1.00 each and 4 softballs worth $2.00 each. What is the total value of the baseballs and softballs on hand?

Often a perplexed student will say, "Oh yes, I see how to do that now," and can proceed from the easier version to solving the original one. You have accomplished your primary objective, having the student analyze the problem and decide how to solve it. Now the student, knowing what to do, can grapple with the more complicated arithmetic in the original.

*Reducing reading difficulties*

We have a tendency to elaborate the story to add realism to the problem. Our aim is to show a wide variety of ways in which mathematics can be used in the real world. Since this elaboration of words can cause difficulty, one way to help a student is to extract the essential information and present it in simpler language:

3. 247 baseballs worth $2.37 each.
   142 softballs worth $3.84 each.
   Worth how much all together?

Version 3 uses eight different words; version 1 uses eighteen different words. Although some of the realism has been sacrificed, the task of the solver remains the same: to analyze the problem and perform the necessary operations. The table below presents four versions of the problem.

*Simplifying Numbers→*

| | | |
|---|---|---|
| | A sporting goods store has 247 baseballs worth $2.37 each and 142 softballs worth $3.84 each. What is the total value of the baseballs and softballs on hand? | A sporting goods store has 3 baseballs worth $1.00 each and 4 softballs worth $2.00 each. What is the total value of the baseballs and softballs on hand? |
| *Simplifying Reading* ↓ | | |
| | 247 baseballs worth $2.37 each. 142 softballs worth $3.84 each. Worth how much all together? | 3 baseballs worth $1.00 each. 4 softballs worth $2.00 each. Worth how much all together? |

The ultimate in reducing reading difficulties is to eliminate them altogether. It is highly recommended that when you are teaching story problems, you read the problem aloud several times or have several good readers from the class do it. When you assign problems for the students to solve independently, you can pair good and poor readers and have both read each problem aloud before they start to solve it.

Another solution is to make a tape recording of a set of story problems. This would take only a few minutes with a cassette tape recorder. You could also add appropriate questions to the tape to assist the student. For example, you might ask on tape:

How *many* baseballs are there?

How *much* does each baseball cost?

How *many* softballs are there?

How *much* does each softball cost?

What *question* does the problem ask?

A student can follow along in the book as the problem is read, replay it as often as necessary, and use the questions to help write down and organize the information needed to solve the problem.

*Capitalizing on students' social context or interest level*

When writing an easy problem for upper elementary grade students, use a "social context" that will fit in with their interests. Even though a seventh grader may only be able to function on the third-grade level in reading and in mathematics, that student's *interests* are the same as those of other seventh graders. Some topics are of interest to a wide range of ages: television, movies, sports, popular music, and school activities.

## Making sure of the given and the goal

Many children plunge into a problem without taking stock of the entire situation. Success is enhanced if the solver has a firm knowledge of the given conditions and of what is to be found. Teachers should ask such questions as—

What do we know for sure?

Anything else?

What are we trying to find out?

What would help us?

How can we find that?

Such questions are posed with two intents. The first is to help the child better understand the problem at hand and the second, to develop the art of self-questioning, which characterizes the problem-solving process. With the end and the beginning in sight, students can be helped in deciding on the steps needed to complete the journey. An all-too-frequent example of an error stemming from failure to take stock of the entire situation would be for the student to "solve" our baseball problem by finding the cost of either the baseballs or the softballs and accepting that as the answer to the problem.

## Acting out and representing problems

One purpose of acting out or representing a problem is to help students understand the problem's applicability to real life. One way of doing this is to change the problem statement. Here, however, we would leave the wording unchanged but provide additional aid to help students visualize the real situation the problem is describing.

*Acting out problems*

Let us begin with our original baseball problem. To inject more reality we might say, "Imagine that you own the store. You need to know the value of your stock on hand, and so somebody has to count it. Probably the person doing the counting will fill in an inventory sheet like this" (fig. 1).

---

### INVENTORY

Total Cost

| | | | |
|---|---|---|---|
| **Baseballs** | Number_____ | $_____ each | _____ |
| **Softballs** | Number_____ | $_____ each | _____ |
| Total cost of baseballs and softballs | | | ======= |

---

Fig. 1

"Now, Doug, can you fill in any of the blanks on this sheet? What numbers are given to you in the problem?" And when that is done, say, "Good. Now, what would you have to do next to fill in the other blanks?"

The inventory sheet is probably pretty close to reality. An employee might be given minimal instructions such as, "We need to take inventory—count the stock and fill in the sheet."

*Representing by a sketch*

When possible, it is often helpful to draw a sketch or a diagram.

John, Alex, and Maryann live on the same road. John lives 10 kilometers from Alex. Maryann lives 2 kilometers from Alex. How far does John live from Maryann?

A typical student's answer to this problem would be either 8 or 12 but not both. Possibly the student has in mind a picture of the relationship of the three houses on the road, but it is more likely that he or she has decided on the operation, guessing it to be either an addition problem or a subtraction problem. A sketch of the problem helps the student to see that there really are two possible answers (fig. 2).

John's house       ‹10 Kilometers›       Alex's house

Where could Maryann's house be?

Fig. 2

Sketches can help children decide on operations and reject impossible answers. Consider this problem:

Ninety-six children are to be placed in rows, with eight children in each row. How many rows will there be?

After sketching three or four rows of children, as shown in figure 3, many children will be able to analyze the problem correctly and, more importantly, experience the benefits of taking a few moments to draw a picture of a problem.

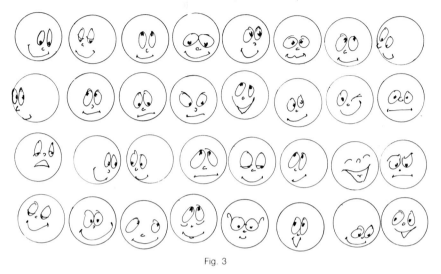

Fig. 3

### Working the problem

So far, we have selected an appropriate problem or created one; we have reduced or avoided reading difficulties; we have helped the students to understand the problem by representing it in some way or by acting out the situation described in the problem. Next we must select and perform the operations needed to work the problem. At this point the students should have time to think and work in a non-competitive atmosphere. Use this time, if you can, to give individual encouragement or hints, to draw a sketch, or to consider all the givens or clearly establish the goal. To avoid time pressure that could cause children to jump to conclusions or copy someone else's work, make the problems due later in the day or have other work available when they finish. Problem solving is not a contest of speed.

### The importance of explaining

It is assumed that we are going to make an active effort to *teach problem solving* rather than just *assign problems*. But even after the careful construction of appropriate problems and efforts to represent them so they are understandable, there will still be some students who ''get it'' and some who don't. It is a good idea to ask one who does to put the solution on the board and explain the work. However, that may not go far enough in helping the student who is having trouble. What that student needs is further explanation that clarifies why the process fits the problem. In the course of this explanation we should also find out, if possible, what it was about the problem that puzzled the student.

Consider again the problem about baseballs and softballs. Suppose it has been solved and the arithmetic displayed on the chalkboard as in figure 4.

| Cost of Baseballs | Cost of Softballs | Total Value |
|---|---|---|
| $2.37 | $3.84 | $585.39—baseballs |
| × 2 47 | × 1 42 | $545.28—softballs |
| 16 59 | 7 68 | $1130.67 |
| 94 8 | 153 6 | |
| 474 | 384 | |
| $585.39 | $545.28 | |

Fig. 4

This is a good example of solving a problem one part at a time. One element of the explanation might go as follows:

"John, we can't find the total value of the balls all at once so we look at the parts of the problem. Just look at the baseballs: 247 of them worth $2.37 each."

Perhaps John has trouble seeing that this is a multiplication problem; if he could see this readily, it is likely he would have solved the problem. We begin to work on this by using the *smaller numbers* strategy:

"Suppose we had two baseballs, each worth $1.00. How much? Suppose we had four baseballs, each worth $1.00. How much? Suppose . . . (several more questions) . . . . Now, John, how are you getting those answers? What operation are you using?"

This last question is crucial to finding out whether the student recognizes the circumstances under which multiplication is the appropriate operation. If, at this stage, John does not see that multiplication is the operation to use, we need to go back and teach him more about multiplication. Let us suppose, however, that John does at this point understand that multiplication is what we need. We continue:

"Yes, John, that's good. You *do* multiply. Now how about the problem where we have 247 baseballs worth $2.37 each. Do that one."

This is the second crucial point. Can John do that multiplication exercise? If not, then of course he can't solve the problem, and, in fact, we had no business assigning it in the first place. It is pointless to assign a story problem for students to work if, even though they can analyze it correctly, they fail to get the answer because they can't do the calculation.

Keep in mind that the primary objective in solving a story problem is to analyze the situation and select the appropriate operations. Since we want to help students be as successful as possible, why not let them use calculators for story-problem activities? The amount of computational practice they would miss would be very small, and computational practice is not the primary objective anyway. In fact, when we adults have any lengthy arithmetic applications, we almost always use a calculator of some kind. Story problems, especially when the numbers get large (as they can in realistic applications), are an ideal place to introduce calculator activities in elementary school.

### Some different problem-solving activities

Many or most of our problem-solving activities employ problems from the basal mathematics text. To provide variety and to help children gain further insight into the problem-solving process, we suggest that on occasion you use one of the following problem-solving activities:

*Problems without numbers*

Write a problem on the chalkboard without the numbers. Begin with version 1 of the baseball problem and delete the numbers:

A sporting goods store has some baseballs and some softballs on hand. A baseball and a softball are not worth the same amount of money. What is the total value of the baseballs and the softballs on hand?

One question for the class is "What would you do to answer the question in this problem?" Since there are no numbers, the students cannot jump in and start calculating; they are forced to develop a plan for solving the problem. Another good question to ask is, "What would you need to know to solve this problem—what information do you need?"

Another type of problem without numbers is one that leads to a function rather than an exact answer:

Tiffany's mother went to the store to buy cans of soup. How much money did she spend?

Again the questions "What would you do to find the answer?" and "What information would you need?" are in order. You may also suggest to the class that they make a table, using $P$ as the price of each can and $C$ as the total cost.

| Number of Cans | Cost |
|:---:|:---:|
| 1 | $C = 1 \times P$ |
| 2 | $C = 2 \times P$ |
| 3 | $C = 3 \times P$ |
| . | . |
| . | . |
| . | . |
| $n$ | $C = n \times P$ |

This activity is extended in graphing (see fig. 5) by deciding on a numerical value for $P$ and graphing the total cost, $C$, as a function of the number of cans.

*Problems without questions*

It is particularly interesting to describe a situation and ask the class to make up questions.

Mary and Sally go to the store. Mary has 58¢ and Sally has 62¢. They want to buy a Frisbee that costs 92¢.

Some questions children might suggest are—

Do they have enough money?

How much change will they get back?

Should they each get the same change?

How much should each girl pay for her share of the Frisbee?

If they plan to share the Frisbee, how much change should each girl get?

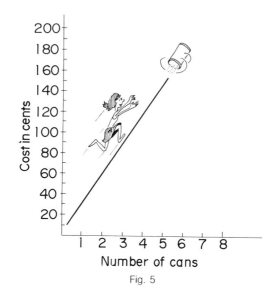

Fig. 5

When the various questions have been written on the board, you have a whole set of problems to solve based on just one situation. The children are active participants; these are *their* problems, not just problems from the book.

*Children's own problems*

For younger children, you can combine a language arts lesson and a mathematics lesson by having children describe their experiences. To make sure you end up with a story problem, give the story direction through leading questions.

"Amanda, do you ever go to the store? What store do you like to go to?"

Amanda went to the Handi-Market.

"What do you like to buy? How much do those things cost?"

She bought candy for 25¢ and gum for 12¢.

You, or Amanda, or another student may ask a question.

How much did Amanda spend?

## Emphasizing Analysis

We have discussed mainly the sort of story problems commonly found in textbooks. These problems are important, and they do illustrate and provide practice in applying mathematics to everyday situations. Most of these problems are solved by performing one or two operations on the numbers given in the problem and obtaining a single exact answer. Other kinds of problems (where the path to the answer is not simply operating on the given numbers) are possible and desirable. In this section we present some problems of the second kind and illustrate the thinking procedures that lead students to solutions. It is interesting to note that several of these problems occur in algebra. Although the formal solution by means of algebra might sometimes be difficult for the average ninth grader, many fifth graders can solve them with great success and enthusiasm.

### Trial and error

Trial and error is not frequently used as a problem-solving procedure in elementary school. Systematic trial and error, however, is a good problem-solving procedure, especially when calculators are available. This strategy places many difficult problems in the range of elementary students. Consider this problem:

I am thinking of two 2-digit numbers. They have the *same digits*. The sum of the *digits* of each number is 10 and the difference between the numbers is 18. What are the numbers?

Examine a systematic trial-and-error approach to the problem.

1. What are all the numbers having *digits* that add up to ten? List them: 19, 28, 37, 55, 64, 73, 82, 91.

2. What differences do we get when we subtract numbers having the *same* digits but with these digits reversed?

$$91 - 19 = 72$$
$$82 - 28 = 54$$
$$73 - 37 = 36$$
$$64 - 46 = 18$$
$$55 - 55 = \phantom{0}0$$

3. The answer to our problem, then, is the pair 64 and 46. We may also ask, "Is this the answer?" and "Is this the only answer?"

Another example of organized trial and error emerges in solving the problem of making a rectangular garden with the largest area one can form from a fence (perimeter) of a given length. Let's take, for example, a fence 20 units in length (fig. 6).

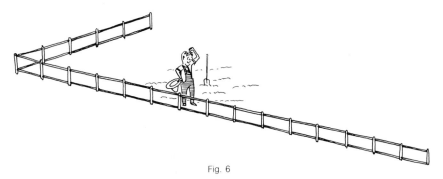

Fig. 6

1. A length of 7 and width of 3 meets the requirement of a perimeter of 20 for our rectangle, and 7 × 3 gives an area of 21 square units (see fig. 7).

2. A length of 8 and a width of 2 also meets our perimeter requirement. But 8 × 2 gives us an area of only 16 square units.

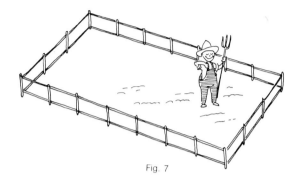

Fig. 7

3. A rectangle of length of 9 and a width of 1 also has a perimeter of 20. But now our area is only 9 × 1, or 9, square units.

4. Hey, we are getting smaller and smaller areas—not the larger one we need!

5. Let's organize what we have so far:

| Length of Rectangle | Width | Area |
|---|---|---|
| 7 | 3 | 21 |
| 8 | 2 | 16 |
| 9 | 1 | 9 |

6. What do we see? Besides the areas getting smaller as the length increases, the length plus width always totals ten! Is this an accident? No—the length added to the width will always total half the perimeter—in this case, 10.

7. What other number pairs can we use? 6 and 4, 5 and 5, and 4 and 6 are possible. What about 6.5 and 3.5, or 5.5 and 4.5? Let's put all these in a table (a good way to organize data) and see what we have. Let's use a calculator here so we don't get bogged down in the computation.

8.

| Length | Width | Area | Length | Width | Area |
|---|---|---|---|---|---|
| 9 | 1 | 9 | 5 | 5 | 25 |
| 8 | 2 | 16 | 4.5 | 5.5 | 24.75 |
| 7 | 3 | 21 | 4 | 6 | 24 |
| 6.5 | 3.5 | 22.75 | 3 | 7 | 21 |
| 6 | 4 | 24 | 2 | 8 | 16 |
| 5.5 | 4.5 | 24.75 | 1 | 9 | 9 |

Examining these results leads us to the conjecture that the *square* (the 5 × 5 rectangle) has the largest area. If our conjecture concerning the square holds, we could repeat the problem for perimeters of 30, 40, and so on.

The point we are making is that *organized* trial and error is a viable technique for solving some mathematics problems. It is a technique that encourages analyzing the problem and the data or the results of our trials as we go along. It is *not* a series of wild guesses.

Organized trial and error can be used on other story problems and, in fact, may help children to select the needed operation. Consider this problem:

Jennifer's dad pays $28 a month for her ballet lessons. If she goes to ballet on Monday and Thursday of each week, *about* how much does each lesson cost?

To begin, we must decide about how many lessons Jennifer has in a month. Examining a calendar would indicate that she has about nine lessons each month. Now let's use organized trial and error. Let's try $2 a lesson. Two dollars for each of nine lessons gives 2 × 9, or $18, a month—not enough. Try $3 a lesson. Result—$27 a month. We're getting closer. Four dollars a lesson gives $36 a month—way too much. A little reflection should indicate that the cost is between $3 and $4 a lesson, but much closer to 3 than to 4. Organizing our work in a table produces figure 8:

$$2 \times 9 = 18$$
$$3 \times 9 = 27$$
$$4 \times 9 = 36$$

Fig. 8

We are looking for

$$\square \times 9 = 28.$$

Look at the results so far. Do they suggest what operation to use on 9 and 28? The trial-and-error approach just described may, indeed, be lengthy if our goal is an answer accurate to the nearest cent—in this example, $3.11. But how important is a precise answer? In real life an exact answer may be unrealistic. After all, Jennifer's dad *never* pays $3.11 for any lesson. Isn't it more important to be able to decide that division is the appropriate operation to apply? And isn't the numerical answer less important than realizing that if one is "stuck," a systematic series of trials can point the way to a correct solution? Organized trial and error is a powerful strategy for solving story problems. It encourages analysis and relieves the pressure to get the correct answer right away.

### Looking back

In their haste to go on to another problem or some other work, teachers and children may fail to look back at the analysis and solution of a problem. This is unfortunate. Capitalize on their feelings of accomplishment to review the procedures used in solving the problem. Stopping when an answer is obtained gives undue emphasis to the importance of answers. Take a few moments after a problem has been solved to reflect on the plan used to solve it. Here is where you can reinforce such techniques as careful reading, drawing pictures, organizing guesses, using easier numbers, and reading carefully to be sure of what is known and what we need to find. After all, these are the very things we want children to do when they meet up with story problems or any application of mathematics. When looking back, be sensitive to other ways to solve the problem. Encourage and reinforce children's efforts to analyze problems. You may find looking back to be your most effective technique in teaching children to solve problems.

Remember also this key to improving attitudes toward problem solving: Frequently give children short sets of problems with which they can experience success.

*10*

# Textbook Problems: Supplementing and Understanding Them

**Jeffrey C. Barnett**
**Larry Sowder**
**Kenneth E. Vos**

THERE is general agreement among researchers and educators that problem solving is a complex process. In an effort to help understand this process, models have been developed which divide problem solving into a number of stages. Common to the beginning stages of most of these models is the fact that students must be motivated to engage in the problem-solving activity, and should have the reading skills needed to understand problem statements.

In the discussion to follow, we should like to consider some ways teachers can help students through the beginning stages of the problem-solving process. First, we shall consider methods of developing student interest in word problems. These methods include ideas for both teacher-generated and learner-generated problems, as well as alternative modes of presenting problem situations. Second, we shall explore some of the language factors that make understanding word problems difficult for many students. Some specific ideas will also be considered for overcoming the difficulty associated with differences between ordinary prose and the language of mathematical word problems.

## Supplementing the Textbook

### Themes that interest students

Students will usually attack most enthusiastically those problems that they find interesting and appealing. They are often more successful with interesting problems

The ideas suggested in this chapter were based on work supported by the National Science Foundation under Grant No. SED77-19157. Any opinions, findings, and conclusions expressed here are those of the authors and do not necessarily reflect the views of the National Science foundation.

than with not-so-interesting ones. Textbooks sometimes provide word problems that have excellent, engaging themes. Since teachers must select from the problems in the textbook and sometimes produce some supplementary ones, they need to ask what themes students find interesting.

The most suitable themes originate from the teacher's familiarity with a particular class and with school or community events. Has the class just completed a field trip to a zoo? Would some of the class members like to own a motorcycle? Was there a television special last evening featuring a singing star? Did it snow? A sizable snowfall might lead to the following problems:

- What is the volume of snow on the school grounds? If we make one big ball out of it, how big would the ball be? How much would it weigh? If all the snow melted into the local swimming pool, would it fill it up?

- How many snows like this one would it take to set a new record? One year it snowed 25.4 meters on a mountain in Washington. How many snows like ours would it take to make that much?

- We get about 90 centimeters of precipitation a year. If all that were in one snowfall, how deep would it be?

- About 3 500 000 cubic meters fell in a big avalanche in the Alps. If all that snow were spread evenly over the football field, how deep would it be?

Encyclopedias or a book of records can supply related data.

Teachers might ask their classes to suggest appealing topics for problems or give the class a list of topics they could use as a springboard for producing problems. One list that interested preadolescents and early adolescents included pets, sports, working with the hands, outdoor games, travel, association with peers, spending money, living outdoors, and watching cartoons and comedies.

Problems that involve familiar settings often result in better performance for average and below-average learners. Steel-mill production, for example, is a less familiar setting for most students than buying candy. "Familiar," however, does not necessarily mean "real life." The charming problems produced in response to the *Arithmetic Teacher's* call for "My Problem-Solving Animal" (Ockenga and Duea 1977) illustrate the appeal of fantasy. (See a ten-year-old's problem in fig. 1.)

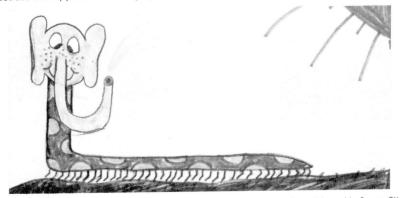

Fig. 1. "I am Elepede-asnake. If I crawl 60 miles a week, how many miles will I crawl in 3 years?"

## Pupil-generated problems

As with "My Problem-Solving Animal," *encourage the children themselves to make up interesting problems.* Pupil-generated problems will usually be of interest to other pupils, and the processes involved in thinking up and solving these problems may improve their performance on other problems. The amount of guidance should vary to suit the sophistication of the children. Sharing problems written by others should be a part of the instructional plan with all learners.

Some examples follow in which the pupils are asked to make up problems.

• Given a picture, the children tell a story and write a number sentence.

*First grade:* Show a sequence of pictures such as that in figure 2. Ask members of the class to tell a story for the pictures. Ask them to write a number sentence that fits the story.

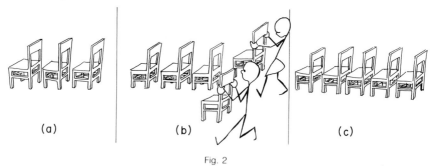

(a)                    (b)                    (c)

Fig. 2

• Given numerical data, the pupils make up problems.

*Fourth grade:* Display a menu (such as that in fig. 3) on the chalkboard or with an overhead projector. Ask each pupil, individually or in pairs, to write and solve a problem based on the menu.

| Menu |
|---|
| Hot dog—50¢ |
| Peanut butter sandwich—35¢ |
| Toasted cheese sandwich—40¢ |
| Tuna fish sandwich—50¢ |
| Apple—20¢ |
| Banana—20¢ |
| White milk—15¢ |
| Chocolate milk—15¢ |

Fig. 3

• Given a theme, the learners make up problems.

*Seventh grade:* Have students search in their science or health books for guidance in making up a problem about heartbeats. Use a few of the problems during class and collect the rest for use by the health teacher. Figure 4 shows some un-edited problems about heartbeats written by seventh graders.

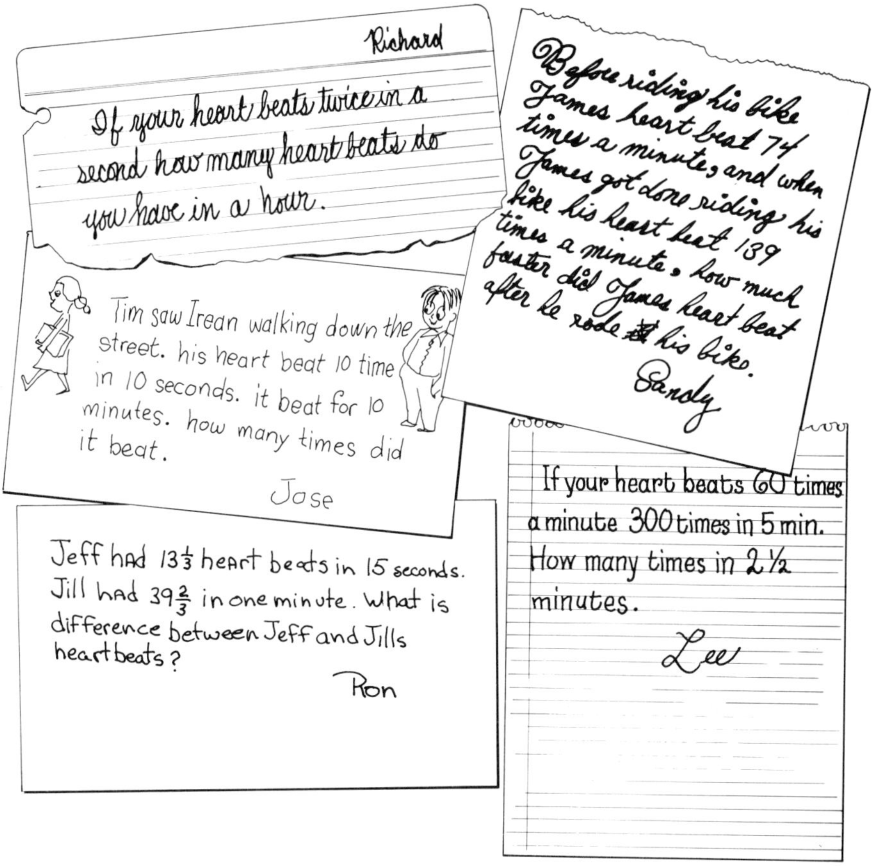

Richard

If your heart beats twice in a second how many heart beats do you have in a hour.

Before riding his bike James heart beat 74 times a minute, and when James got done riding his bike his heart beat 139 times a minute. how much faster did James heart beat after he rode his bike.

Sandy

Tim saw Irean walking down the street. his heart beat 10 time in 10 seconds. it beat for 10 minutes. how many times did it beat.

Jose

If your heart beats 60 times a minute 300 times in 5 min. How many times in 2 ½ minutes.

Lee

Jeff hAd $13\frac{1}{3}$ heArt beats in 15 seconds. Jill hAd $39\frac{2}{3}$ in one minute. What is difference between Jeff and Jills heartbeats?

Ron

Fig. 4

Such activities provide a change of pace for the class. Student-generated problems frequently include extraneous data and collectively may involve more than one operation or process—thus they *must* be read. Furthermore, since such problems are ''theirs'' and often feature classmates, the students *want* to read and solve them.

### Alternatives to words

Textbook problems are usually presented in verbal form. Sometimes a problem is accompanied by a picture involving a theme, or data presented in a picture, map, chart, or graph. If your particular mathematics textbook does not include any maps, tables, or graphs, you might wish to collect some, since they are important in the overall problem-solving process.

Problems that present key data through pictures or drawings with a minimum use of words (fig. 5) are important for improving problem-solving ability. Newspaper advertisements, television commercials, and catalogs, which usually present their messages largely through pictures, can be good sources of problems. Picture problems are of help to the poor reader or the student who is more visually oriented. It is essential that drawings be accurate. Problems based on inaccurate drawings (see fig. 6) usually lead to difficulties.

How much for 3 ?

Fig. 5

Who is taller? By how much?

Fig. 6

Actual objects can also be used to present problems. Such projects as the play grocery store, which is a common activity project in the primary classroom, offer many opportunities for designing problems that do not depend entirely on written words. Real objects can also be used in more advanced mathematics classes. Ames's article (1977) shows how a bicycle can be used in a class to create problems about ratio, circumference, or distance and speed. Since measurement is the source of so many real-life uses of numbers, if is desirable to incorporate the measurement process into problems (fig. 7).

Find the average mass
of the rocks.

Fig. 7

These "alternative to words" lessons should occur often, since some children have trouble adjusting to the unexpected in the classroom. Problems presented with pictures or concrete materials create greater interest, involve an unusual degree of realism, and perhaps assist some students who have special difficulty with word problems.

Word problems are important, however, and deserve much attention. Since many difficulties with word problems can be traced to poor language skills, some ways of dealing with these language-based difficulties will be discussed.

## Overcoming Language Barriers

The role of language in the problem-solving process has received considerable attention as an area of research for many years. Unfortunately, there is little evidence that special attention is devoted to reading problems and language processing in mathematical problem solving in the classroom, especially at the junior and senior high school levels. There is ample evidence that reading and language processing are crucial abilities that influence problem-solving behavior during the early stages of the problem-solving process. There is further evidence that special instruction in reading mathematics problems can have a positive effect on success in problem solving.

### Why the barriers?

Differences between reading prose and reading mathematical word problems have been noted for several years by language arts and mathematics education researchers such as Earp (1970) and Henney (1971). Perhaps a starting point in designing reading instruction in mathematics is to convince teachers and students that these differences do exist and do call for different approaches. Some of these differences are noted here.

*Density.*

Mathematical word problems are more compact and conceptually dense than ordinary prose. An ordinary paragraph of prose usually contains one major idea, but mathematical word problems often squeeze several important ideas into a single sentence:

> To raise money for new playground equipment, Mrs. Maple's fifth-grade class sold 180 boxes of candy at $1.50 a box and 40 T-shirts at $2.00 each. If each box of candy and each T-shirt costs the class $1.20 and the students wish to award $3.00 in prize money, how much profit did the class make on the sale?

Teachers can help their pupils cope with this concentration of information by suggesting a slower reading rate with increased attention to detail and to the relationships within the problem statement.

*Thought units*

The writing style found in word problems is usually different from that used in most other types of prose. Word problems usually contain relatively short thought units that are closely related to each other. A single problem statement often contains many such thought units. The teacher can suggest a variety of methods—pictures, charts, and graphs, for instance—for keeping a record of the data. In the following problem, the teacher might suggest a graph (fig. 8) to help students organize the data.

On a recent field trip, 3 children collected 5 or fewer specimens of insects each, 7 children collected between 6 and 10 specimens each, 6 children collected 11 to 15 specimens each, 8 children collected between 16 and 20 specimens each, and 2 children collected over 30 specimens each. What part of the class collected fewer than 16 specimens? What part of the class collected more than 10 specimens?

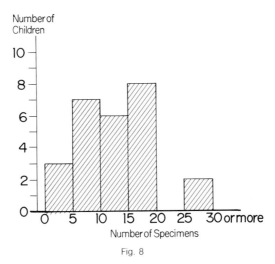

Fig. 8

*Context clues*

Word problems often lack the relatively rich context clues typically found in English prose. This makes words and phrases difficult to identify. Adjectives are usually more important in word problems than in ordinary prose, since they help distinguish between important variables or indicate relative magnitudes that must be taken into consideration in solving the problem. In the following example, the role of adjectives is apparent:

The large, gray ape is twice as heavy as the smaller one, which weighs 15 kg more than the brown orangutang. The brown orangutang's weight is 7/8 of the weight of the red orangutang, which is 20 kg less than the weight of his keeper. If the keeper weighs 100 kg, how much does the heaviest ape weigh?

In the absence of context clues in other problems, children need systematic practice in interpreting the action of the problem and in translating key verbs into the appropriate mathematical operations.

*Vocabulary differences*

Another area of difficulty in solving word problems relates to vocabulary. Often the meaning of a word in a mathematical problem is entirely different from the meaning of that same word in ordinary prose. The words *operation, times, mean, altitude, base, left, power, by, range, degree,* and *prime* often have different meanings in mathematical problems. Note the differences in meaning of the italicized words in the following examples:

Prose: At *times,* the *power* of John's *left* arm was amazing as he threw his fast ball.

Word problem: If Mary subtracts 9 from 3 *times* the second *power* of 4, she will be *left* with what number as an answer?

Prose: The *altitude* of the airplane was 30 000 feet.

Word problem: The *altitude* from vertex *A* of equilateral triangle *ABC* is 10 meters. How long is each side of the triangle *ABC*?

New terms, such as *denominator, isosceles, hypotenuse,* and *sine,* may have no relationship at all to the learner's everyday vocabulary and therefore must be learned outside any familiar verbal context.

*Continuity*

Prose usually possesses a continuity of subject and ideas from sentence to sentence and paragraph to paragraph. There is often little continuity among word problems in any given set. When problem sets are constructed to be similar in form, students often develop a mind set and begin to process each problem in the same way. Then when a different sort of problem appears, they have difficulty adjusting to the differences in language and sequence of information. This difficulty can usually be lessened by exposing them to a variety of problems in terms of context, content, and syntactic structure.

*Reading patterns*

Normal reading patterns are often ineffective for word problems. Symbols and numerals in word problems can cause breaks in the student's chain of thought. In problems such as the following, the reader may focus on the numbers in the problem statement and become distracted from the relationships implied by verbs and nouns.

During the week of her birthday, Judy received in the mail 2 one-dollar bills from her brother, 3 five-dollar bills from her Aunt Sally, one ten-dollar bill from her Uncle Ned, and a bill from the bookstore for $3.00. How much money did she receive in the mail for her birthday?

Familiar reading patterns are particularly disturbed when a picture, graph, or chart accompanies the problem statement, requiring the learner to relate portions of the problem statement to the accompanying visual before completing the reading of the problem.

*Adjusting reading habits*

Earp (1969) has suggested that differences in English prose and mathematical word problems require several reading adjustments for mathematical material: (1) a slower reading rate; (2) varied eye movements, including some regressions: (3) an attitude of aggressiveness and thoroughness. The teacher can help students make these three adjustments by providing practice, pointing out differences between reading patterns that are effective for each type of material, providing motivational problems that are interesting to read and sequenced to provide a balance between success and challenge, and—perhaps most important—providing reinforcement for positive attitudes toward problem solving.

*Coping with confusion over vocabulary*

The student must be able to interpret words in a mathematical context in order to understand the problem clearly. This ability is not always easy to cultivate but can be developed with practice. Some words—such as *sum, total, decrease,* and *difference*—can provide clues to help learners translate the data contained in the problem statement into equations, as in the following example:

> The sum of 8 and 5, decreased by the difference between 12 and 7, gives a total of how many?

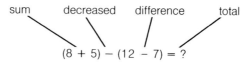

Unfortunately, the same term may not always indicate the same operation or procedure in different contexts.

One productive method for teaching the relationship between key words and the underlying mathematical structure of a word problem is to have students underline the words they consider clues to the operations required. For example, in the problem below, the word *and* indicates the operation of addition, the word *of* indicates multiplication, and the word *left* indicates subtraction.

> Mary purchased 12 tomato plants on Friday <u>and</u> 6 on Saturday. She gave 1/3 <u>of</u> them to her brother. How many plants did she have <u>left</u>?

$$\text{``of''} \quad \text{``and''}$$
$$\downarrow \qquad \downarrow$$
$$\frac{1}{3} \times (12 + 6) = \frac{1}{3} \times 18 = 6$$
$$\text{``left''}$$
$$\downarrow$$
$$18 - 6 = 12$$

Although such words are often indicators of required operations, students should be cautioned to look carefully at the context, for numerous exceptions do exist. For example, in the first of the three problems that follow, the word *and* does not imply addition. In the second, the word *left* does not imply subtraction. The word *sum* in the third problem does indicate addition, but the operation has already been performed and therefore is not required in the solution process.

- What is the product of 5 <u>and</u> 7?
- John walked 8 blocks north and turned <u>left</u>. He then walked 6 blocks east. How far was he from his original starting place?
- Judy found that the <u>sum</u> of 5 test scores was 432. What was the average score?

Perhaps the best method is to have children look for *potential* key words as they read through the problem. During a second reading they can use contextual clues to help determine which of the identified words are actually operational or procedural indicators. Systematic exercises of this type can help them focus on semantic hints to discover which operations or procedures are required for a solution.

Teachers can help their pupils read and understand word problems by devoting special attention to new and unfamiliar vocabulary. Special attention should be given to entire word families—different forms of the same basic word. There is evidence that marked differences may exist between a child's familiarity with one word and a different form of the same word. For example, Kane, Byrne, and Hater (1974, pp. 75–90) found that 76.6 percent of seventh- and eighth-grade children were familiar with the term *associative* but that only 39.1 percent were familiar with the term *associativity*. Some teachers begin each new chapter of the textbook with a pretest of the new mathematical terms used in the chapter. It is advantageous to have students read word problems aloud to determine which words and phrases are causing the most difficulty. New mathematical vocabulary should be displayed on the chalkboard or overhead projector, pronounced, defined, and used in an appropriate context. Students should be encouraged to write and solve word problems that contain the new terms and demonstrate that they understand their meaning. Familiar words with special mathematical meanings should be emphasized by writing word problems and ordinary prose to illustrate how the meanings differ.

*Setting reading goals*

Students are given narrative materials to read for a variety of reasons. It may be for enjoyment or to obtain information about a particular topic, but in any case the child is reading for a single purpose. Interestingly, researchers have noted that schoolchildren seldom read mathematical word problems with just one purpose in mind (Spencer and Russell 1960). For example, they may read a problem to get an overall view of the idea of the task, to note action sequences indicated by verbs and the position of data, to relate ideas, or to determine the question asked and the required form of the answer. Solving a particular problem may require reading and re-reading the problem statement several times to achieve these purposes. The teacher can help by encouraging students to reread problems for specific purposes.

The following procedures have been shown to be effective in assisting learners to process language and data contained in word problems (Earp 1970; Barnett 1974):

▶ *Reading the problem statement through completely to obtain a general idea of the setting and to visualize the situation.* To help pupils avoid centering on the numerals too early, the teacher could ask them to rewrite the problem without numerals:

*With numbers:* Mary drives north for 20 minutes at 90 km an hour, and then drives east for ½ hour at 80 km an hour. If she drives in a straight line back to where she started, at what speed must she travel to reach home in 45 minutes?

*Without numbers:* Mary drives north at one rate, and then east at another rate. If she drives straight back to where she started, at what rate must she travel to reach home in a given time?

The second version is one of several ways of rewriting the problem without the numerals. The numberless version could clarify the overall situation and help students
see this as a rate problem involving the Pythagorean theorem.

▶ *Rereading the problem statement to understand the facts and relationships.* Adjectives sometimes have a special role:

> The brown horse can run 5 km an hour faster than the black horse, which can run
> 10 km an hour faster than the old gray mare. If the old gray mare can run at 11
> km an hour, how fast can the brown horse run?

In this example, the color adjectives are necessary to distinguish one horse from another. Although the adjective *old* is not strictly necessary, it does reinforce the fact
that the gray mare is the slowest horse.

Problems containing many pronouns can be confusing. During the rereading, pupils should be encouraged to substitute nouns for pronouns to make sure they understand the problem.

> Mary's dog had a puppy which ~~she~~ named Spot. ~~She~~ weighed three pounds at
> birth. Mary observed that ~~she~~ gained two pounds every four weeks. At that rate,
> how many pounds will ~~she~~ weigh at the end of ~~her~~ first year?

(*Mary*, *Spot*, *Spot*, *Spot*, *Spot's* handwritten above the crossed-out pronouns)

Children sometimes obtain the correct answer to the wrong question—that is,
they do not answer the question that is asked. During this rereading, teachers should
ask pupils to write the form of the answer, leaving a blank for the numerals as illustrated in the following example:

> A box-shaped container is 35 cm by 67 cm by 81 cm. How many cubic deci-
> meters of water will it hold?
>
>                    Answer =_____cubic decimeters

▶ *Scanning the problem statement to note difficult or unfamiliar vocabulary or
concepts.* Some suggestions for helping learners understand unknown vocabulary
terms have already been discussed.

▶ *Rereading to help organize the steps leading to a possible solution.* During this
reading, teachers may encourage their pupils to focus attention on the action of the
problem, centering on important verbs. This is also a good time to identify irrelevant
data (distractors), particularly if the problem is lengthy and seems to contain a great
deal of information. In the following example, the student has been instructed to
cross out all unnecessary information to make the problem simpler:

> ~~John's aunt~~ Mary uses two ~~3-pound~~ sacks of ~~enriched~~ flour to make eight ~~1-
> pound~~ loaves of ~~rye~~ bread. How many ~~1-pound~~ loaves can she make with five
> ~~3-pound~~ sacks of flour?

Just as many problems contain unnecessary information, others involve important but "hidden" information that is assumed to be known and therefore not explicitly stated in the problem:

Mary spends $1.39 for 250 grams of candy. At the same rate, how much would she have to spend for a kilogram?

Even though the number of grams in a kilogram is not stated, this information is essential to reaching a solution. Teachers should be aware that some errors are due to the lack of knowledge of hidden information rather than to the lack of knowledge of the problem-solving processes themselves.

As a final check before computation, the learner should attempt to estimate the answer to check its reasonableness with the data and action sequences of the problem statement.

▶ *Rereading the problem one more time to check the procedures used and to be sure the solution is complete and in the proper form.*

These five steps are not the only ways teachers can help their students interpret problem statements, but they do form a basis for improved instruction that is supported by research. By helping children realize that procedures for reading mathematical word problems are different from those for reading ordinary prose, teachers can begin systematic instruction and practice on those reading skills necessary for increasing problem-solving ability.

## Summary

We have suggested some ideas that teachers can use to help students begin the problem-solving process. Teaching children to become better problem solvers is not an easy task. However, the job becomes less difficult when they are given the opportunity to work problems that interest them.

It is probably not possible to assess the relative importance of reading ability to the problem-solving process. It is clear, however, that many children with poor language skills never reach the stage of understanding the problem. By systematically providing experiences to help them develop language-processing skills in the area of mathematics, the teacher can help children improve their ability in this crucial area of problem solving.

### REFERENCES

Ames, Pamela. "Bring a Bike to Class." *Arithmetic Teacher* 25 (November 1977): 50–53.

Barnett, Jeffrey C. "Toward a Theory of Sequencing: Study 3–7: An Investigation of the Relationships of Structural Variables, Instruction, and Difficulty in Verbal, Arithmetic Problem Solving." (Doctoral dissertation, Pennsylvania State University, 1974.) *Dissertation Abstracts* 36A (1975): 99–100. (University Microfilms No. 75-15787)

Earp, N. Wesley. "Procedures for Teaching Reading in Mathematics." *Arithmetic Teacher* 17 (November 1970): 575–79.

Henney, Maribeth. "Improving Mathematics Verbal Problem-Solving Ability through Reading Instruction." *Arithmetic Teacher* 18 (April 1971): 223–29.

Kane, Robert B., Mary Ann Byrne, and Mary A. Hater. *Helping Children Read Mathematics.* New York: American Book Co., 1974.

Ockenga, Earl, and Joan Duea. "IDEAS." *Arithmetic Teacher* 25 (November 1977): 28–32.

Spencer, Peter L., and David H. Russell. "Reading in Arithmetic." In *Instruction in Arithmetic,* Twenty-fifth Yearbook of the National Council of Teachers of Mathematics, pp. 202–23. Washington, D.C.: The Council, 1960.

# 11

# Teaching Problem Solving in the Elementary School

John F. LeBlanc
Linda Proudfit
Ian J. Putt

**D**EVELOPING skill in problem solving has long been recognized as one of the important goals in the elementary school mathematics program. Instruction in problem solving has also been recognized as being a difficult task. One reason for this difficulty is that problem solving is a complex process rather than a set of simple algorithmic skills. Increased awareness of the importance of problem solving and of the difficulty of problem-solving instruction has resulted in increased efforts to identify specific instructional techniques for teaching problem-solving skills. Some specific suggestions for improving problem-solving instruction are made in this essay.

## Two Types of Mathematical Problems

The standard textbook problem and the process problem are among the types of mathematical problems found in the elementary school mathematics curriculum. A description of each is given, and some reasons for its inclusion in the curriculum are discussed.

## Standard textbook problems

The most common type of textbook problem introduces or follows the development of an arithmetic operation such as the multiplication of whole numbers. A characteristic of the standard textbook problem is that it can be solved by the direct application of one or more previously learned algorithms. The basic task is to identify which operations or algorithms are appropriate for solving the problem. Standard textbook problems allow children to work with the operations in a concrete or real-world context. The purposes of standard textbook problems include improving the recall of basic facts, strengthening skills with the fundamental operations algorithms, and reinforcing the relationship between the operations and their applications in real-world situations.

The problem situation is normally presented using pictures, short phrases or sentences, paragraphs, or a combination of these modes. In the first and second grades, pictures alone or pictures and words (rebus format) are commonly used to present the situation. In the middle grades fewer pictures are used, and abbreviated story problems often occur. These problems are presented in short phrases and sentences with a minimum of situational information. An example of an abbreviated story problem follows:

> 3 cartons.
> 6 bottles in each carton.
> How many bottles in all?

In grades 5 through 8, the predominant format for the story problem is the paragraph, which gives a more complete description of the situation. For example, the abbreviated story problem above could be expressed in paragraph form as follows:

> Joe went to the store to buy 3 cartons of Coke. If each carton contains 6 bottles, how many bottles of Coke did Joe buy?

## Process problems

The "process problem" is another type of problem that is beginning to appear in textbooks. Whereas standard textbook problems require only the application of operations or algorithms, process problems require the use of strategies or some non-algorithmic approach. An existing algorithm may solve the process problem; however, it would not be available to the elementary school child. This type of problem stresses the process of obtaining the solution rather than the solution itself. Success in solving the problem does not depend on the application of specific mathematical concepts, formulas, or algorithms; rather, the solution requires the use of one or more strategies. Process problems frequently have more than one answer.

Process problems are used to encourage the development and practice of problem-solving strategies. In addition, they provide an opportunity for students to devise creative methods of solution, to share their methods with other students, and to build confidence in solving problems. Process problems also allow students to enjoy mathematical problem solving.

Here is one example of a process problem along with some strategies that might be used to solve it:

> There were 8 people at a party. If each person shook hands with everyone else, how many handshakes were there in all?

*Strategies:*

1. Drawing a diagram

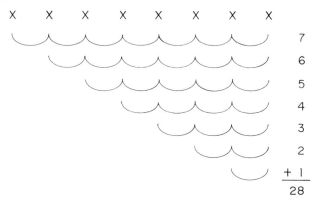

2. Acting out the problem

   Some children may choose to have eight people shake hands with each other, counting as this is done.

3. Making a list

   Other children may give names to the eight people and list the hand-shakes, for example—

| Steve | Diane | Jeff | Eric | Mary | José | Sally | Susan |
|-------|-------|------|------|------|------|-------|-------|
| Diane | Jeff  | Eric | Mary | José | Sally | Susan | |
| Jeff  | Eric  | Mary | José | Sally | Susan | | |
| Eric  | Mary  | José | Sally | Susan | | | |
| Mary  | José  | Sally | Susan | | | | |
| José  | Sally | Susan | | | | | |
| Sally | Susan | | | | | | |
| Susan | | | | | | | |

Strategies used in solving mathematical process problems include making a list, making an organized list using guess-and-test, making a diagram, making a table, using an equation, computing, simplifying, finding a pattern, using deduction, acting out, experimenting, and working backward. Some of these strategies will be used later to exemplify specific instructional moves in teaching problem solving.

## Selecting Problems for Classroom Instruction

Teachers want their students to be successful and confident in solving problems. Motivation is one of many factors that contribute to success and confidence. A child who does not want to solve a problem probably will not solve it! Teachers must select or devise problems that interest the students. A climate of relevance and enjoyment is critical for success in problem-solving instruction.

Selecting problems with an appropriate level of difficulty is also essential. What factors make a problem difficult? Four general factors affect the difficulty level of both textbook and process problems—the choice of vocabulary, the length and structure of the phrases or sentences, the size and complexity of the numbers, and the problem setting, or representation.

Vocabulary should be selected to communicate as simply as possible. Mathematical terms (*perpendicular, multiple,* etc.) should not be avoided, but they must be clearly understood by the student. Reading difficulty increases in proportion to the length and complexity of phrases and sentences in the problem. Problems should be examined to see if a long sentence can be split into two or more sentences or rewritten into a substantially shorter sentence.

Replacing large, complex numbers with smaller, simple numbers allows the student to focus on problem solving rather than on computation. The use of the hand calculator also reduces the difficulty level of a problem having large or complex numbers.

Changing the problem setting, or representation, can alter the difficulty level of a problem. For example, the following is the same basic problem presented in two settings:

How many handshakes would there be if 8 people at a party all shook hands with each other?

How many line segments would there be if each of 8 points was connected with each of the other points?

Other factors affect the difficulty level of textbook or process problems more specifically. These factors are discussed in the sections that follow.

### Standard textbook problems

A usual objective of standard textbook problems requires the child to translate a real-world situation into mathematical symbols so that a previously learned algorithm can be used to solve the problem. Usually, the more complex the algorithm, the more difficult the problem. Thus, a problem involving percentage usually presents more difficulty than one involving simple multiplication. The level of the mathematics or the complexity of the algorithm is a factor in the difficulty of standard textbook problems.

The number of steps is another factor in the difficulty level of the problem. A two-step problem is more difficult than a one-step problem, assuming that other factors related to problem difficulty are constant. Multistep problems are prevalent in upper-grade textbooks and often occur when teachers use problems drawn from real-world or classroom situations. The following problem is a multistep problem:

Mary bought 6 hamburgers at $0.79 each, 4 packs of french fries at $0.45 each, and 6 Cokes at $0.35 each for her friends. How much did she spend? How much would she receive in change from $10.00?

Students may find that the use of the hand calculator facilitates the solution process when the solution requires more than a single operational algorithm.

### Process problems

Process problems vary in difficulty according to the number of conditions that must be satisfied simultaneously. For example, problem A below has three conditions to satisfy, whereas problem B has only two.

> **Problem A.** Jesse's mother paid him $1.60 allowance in quarters, dimes, and nickels. He received 17 coins in all. How many of each coin could his mother have given him?

> **Problem B.** Jane saw 18 chickens and pigs in a farmyard. If she counted 50 legs, how many chickens and how many pigs were in the farmyard?

In problem A the three conditions to be satisfied are (1) there must be 17 coins; (2) there must be at least one each of quarters, dimes, and nickels; and (3) the total value must be $1.60. Satisfying one or two of these conditions presents less difficulty than satisfying all three. In problem B the two conditions to be satisfied are (1) there must be 18 animals, and (2) the total number of legs must be 50. Students often will find an answer that satisfies one condition while ignoring the other(s). For example, the answer "17 pigs and 1 chicken" satisfies the first condition but not the second. Encourage students to examine the sets of answers that satisfy each condition separately. The intersection of these sets is the solution.

The complexity of each condition is another factor affecting the difficulty level of process problems. For example, the difficulty level of problem A would be changed by stating that Jesse received only quarters and nickels, or that he received half-dollars, quarters, dimes, and nickels totaling $2.75. Similarly, the difficulty level of problem B would be changed by stating that there were 18 animals including chickens, pigs, and cows.

Another difficulty factor is the type of strategy with which the problem might be readily solved. For example, children generally find the guess-and-test strategy easier to use than organized lists. More research must be carried out to determine which strategies students find easiest to learn and use. The factor of multiple solutions is related to the strategy factor. Problems having only one solution may or may not be more difficult than problems having multiple solutions. However, if a student is asked to find *all* solutions, a natural question is "How can I be sure that I have all the solutions?" The effect of multiple solutions on the difficulty of the problem needs to be further researched.

# An Instructional Model
# for Problem Solving

Teaching problem solving is a clearly recognized task for the teacher. Yet, this task is difficult compared to teaching mathematical skills or concepts. In problem solving, an individually acquired set of processes is brought to bear on the situation by the problem solver. Helping the child acquire and apply these processes is a more complicated and less well defined instructional task than helping a child learn a computational skill or understand a concept.

Polya, in his book *How to Solve It,* discusses in some detail four phases of problem solving. In the first phase the problem solver must at least understand the question and want to answer it. The person must recognize what is known, what is unknown, and what conditions are present. In the second phase a problem solver might search past experience for a related problem that has already been solved or might tentatively try a number of attacks before settling on one that seems promising. In the third phase the problem solver carries through the plan to a solution or, reaching an impasse, returns to the planning phase. Finally, the problem solver checks the solution against the data and conditions presented in the problem.

These four phases are not necessarily completed in the order listed. For example, a problem solver could reach a solution only to find in the looking-back phase that a condition had been overlooked. Whatever the order an individual problem solver might follow, the phases as presented by Polya have gained wide acceptance as a conceptual model for problem solving. Examples A and B illustrate how the model can be adapted for classroom instruction.

### Example A (standard textbook problem)

Last summer 284 children attended camp each day. A case of cola holds 24 bottles. How many cases of cola were needed each day so that every child could have had a bottle of cola for lunch?

*Understanding the problem*

The teacher might ask questions like these to help the child understand the problem:

Mary, how many children were at camp each day?

Joe, do all the students at camp drink cola for lunch? If not, would that make a difference in our problem?

Manuel, do we usually buy cases of cola when we have a large number of people involved?

If one case of cola was bought, how many bottles would that be?

Sophia, can you tell in your own words what the problem asks you to find?

Furthermore, the children can learn from similar kinds of experiences how to help themselves understand a problem. Children should be encouraged to ask questions for understanding that can be shared with the class as a whole. If the children work in small groups, members of a group might formulate questions for understanding and record these on paper for the group's use. Children may retell the story in their own words as a way of better understanding the problem. These experiences develop children's ability to focus on important information in problems. Students should eventually be trained to ask themselves questions when they are confronted by a problem.

*Devising a plan*

In this phase the teacher should direct children's attention to related problems and previously used strategies where possible. The teacher should also encourage children to share their own strategies with the class. Initially, however, the teacher may have to suggest some strategies to help children get started on planning a solution. One such question might be "Has anyone solved a problem like this one before?" If the answer is yes, the teacher can then ask that child to describe the previous experience to the class. This may help other class members to recall a similar problem. As an alternative when children do not recognize the given problem, the teacher can pose questions related to a simpler problem. The following questions would be suitable:

If there were 72 children at the camp, how many cases of cola are needed daily if each child has a bottle for lunch?

If there were 90 children at the camp, how many cases of cola are needed daily to give each child a bottle at lunchtime?

As the children gain experience in using strategies, the teacher can encourage them to suggest the strategies that they would use to solve the problem.

*Teacher:* How might we make a plan to solve this problem?

*Jenny:* We could keep on adding—24 plus 24 plus 24 and so on—until we get to the first number greater than 284. Then we could count up the number of 24s we added, and that would be the number of cases needed each day.

*Teacher:* That's good, Jenny. You have suggested *repeated addition* as a way of solving the problem. Can someone else suggest another way we could plan to solve this problem?

*Mike:* Instead of adding like Jenny did, I would start with 1 case, which is 24 bottles; so 10 cases is 240 bottles. I would then add 1 times 24, 2 times 24, 3 times 24, and so on, until I found a number equal to 284 or a bit bigger than 284. Then I would add 10 and 2 or 3 to get the number of cases.

*Teacher:* That sounds like a good suggestion, Mike. You would be using *multiplication* to solve the problem.

*Emmanuel:* I would get 284 chips and put them into groups of 24.

*Teacher:* Good, Emmanuel. You would be conducting an *experiment* to help you solve the problem.

*Tania:* The quickest way would be to divide 284 by 24 to find the number of cases.

*Teacher:* Thank you, Tania. You would be using *direct computation* as your strategy.

A discussion of possible strategies may help some students gain a clearer understanding of the problem. The teacher's labeling of the suggested procedure or strategy also improves children's communication skills.

*Carrying out the plan*

The plan selected in phase two is carried out now. In this step children should be encouraged to solve the problem on their own. If the selected plan does not work, the teacher can encourage the child or group of children to try an alternative plan suggested in phase two. The teacher can also delve into the child's understanding of the problem if there is evidence of incorrect or incomplete understanding.

There is a danger here that the teacher will stress this step out of proportion to the other equally important steps in the process. Although computational accuracy is desirable, the teacher should be careful not to stress it so heavily that children tend to equate problem solving with computation. The important aspect of this phase is the child's ability to implement the plan or plans selected in phase two.

*Looking back*

This fourth phase is essential for consolidating the knowledge gained from the solution and for developing in children the processes needed for solving problems. Therefore, it should not be omitted from the instructional sequence. The teacher can ask children to describe to the class the strategy used in solving a problem, and the name of the strategy should be drawn to the attention of all students. A number of different strategies for each problem should be shared, appropriately labeled, and emphasized.

Another aspect of this phase is "looking forward," or extending the problem. The teacher can extend the problem by asking additional questions:

A case of cola has four cartons containing six bottles in each. How many cartons of cola would be needed daily to give each child a bottle for lunch?

If a bottle of cola cost $0.15, how much will it cost each day to give all the children a bottle for lunch? How much change will be left from $50.00?

This standard textbook problem can be most easily and efficiently solved by the direct application of the long division algorithm. Indeed, standard textbook problems most often require an algorithmic approach, but the teacher should be aware of, and sensitive to, solution strategies other than the most obvious and efficient ones. Students should be encouraged to use a number of strategies in solving a particular problem in order to broaden their problem-solving processes. Students can gain confidence in their problem-solving ability by increasing their repertoire and power in using a variety of strategies.

### Example B (process problem)

Susan wanted to buy a candy bar that cost 25 cents. The machine would take pennies, nickels, and dimes in any combination. List the different coins she could use to pay for her candy.

*Understanding the problem*

Some questions that were asked by a teacher or formulated by some fifth-grade students follow:

How much does a candy bar cost? Which coins does the machine take? Can all the coins be the same? Can Susan pay with a quarter? Do you think there is more than one answer for the problem? Can you tell in your own words what the problem is asking you to find?

*Devising a plan*

When this problem was given to fifth-grade students, some suggested they would make a table like the one below:

| Nickels | Dimes | Pennies |
| --- | --- | --- |
|  |  |  |

Others suggested writing down the denominations, for instance, 2 dimes and 1 nickel, 25 pennies. The teacher labeled these strategies *making a table* and *making a list* and wrote these strategies on the board. The teacher did not have to suggest a strategy, but for some problems it may be necessary, initially, to supply a strategy. The teacher was careful, however, to label the strategy to ensure clear communication among the children.

*Carrying out the plan*

Once plans are suggested (phase two), children usually embark on the solution without much trouble. Most fifth-grade children can get some entries in the list, but they tend to end up with incomplete lists because the entries are not organized in any way. If children work in small groups of three or four, the sharing and discussion of answers may lead to a complete solution.

*Looking back*

After solutions to the problem are obtained, the class discussion should focus on having the children introspect—that is, analyze their own strategies and consider alternative strategies.

*Teacher:* John, would you show the class how your group solved the problem?

*John:* We wrote down all the answers we could find, like this:

> 3 nickels and 1 dime
> 25 pennies
> 2 nickels, 5 pennies, and 1 dime
> 4 nickels and 5 pennies

That's all we could find.

*Teacher:* Can you name the strategy your group used, John?

*John:* We just called it "listing."

*Teacher:* Did another group use a different strategy to help them solve the problem? Becky, would you show the class how your group solved the problem, please?

*Becky:* We also made a list by writing the names of the three types of coins at the top of three columns.

| Nickels | Pennies | Dimes |
|---------|---------|-------|
| 5 | | |
| | 25 | |
| | 5 | 2 |
| 4 | 5 | |
| 3 | | 1 |
| 2 | 5 | 1 |
| 1 | 10 | 1 |
| 1 | 20 | |
| 1 | | 2 |
| 2 | 15 | |
| 3 | 10 | |
| | 15 | 1 |

*Teacher:* Are you sure you have all the answers to this problem?

*Becky:* I think so.

*Joel:* Our group had the same answers.

*Teacher:* How can we be sure that we have all the solutions to this problem?

[*No response*] Is there some way we can write down all the answers that Becky's group worked out without missing any of them?

*Stephanie:* We could find out the largest number of dimes we can have, then start with them and work down to the smallest number of dimes.

*Teacher:* Would you draw that on the board so that everyone can see what you mean, please, Stephanie?

*Stephanie:* I would make columns like Becky's group did, but with dimes first, then nickels, and then pennies. See, like this:

| Dimes | Nickels | Pennies |
|-------|---------|---------|
| 2 | 1 | 0 |
| 2 | 0 | 5 |
| 1 | 3 | 0 |
| 1 | 2 | 5 |
| 1 | 1 | 10 |
| 1 | 0 | 15 |
|   | 5 | 0 |
|   | 4 | 5 |
|   | 3 | 10 |
|   | 2 | 15 |
|   | 1 | 20 |
|   |   | 25 |

When you do it this way, there are no more answers to put in.

*Teacher:* That's very good. When we draw up a list like Stephanie's and put the numbers in some order, we call it an *organized list*. I'll write that here beside Stephanie's solution.

The teacher can continue the discussion until the strategies of all the different groups have been examined. As a student outlines another strategy, the teacher encourages the naming of it, if possible. Otherwise, the teacher should supply the appropriate label. Drawing attention to different strategies and their names facilitates communication and aids students in selecting strategies in the future.

To extend the problem, the teacher should ask children questions like the following:

If the candy bar costs 30 cents, show the ways Susan could put coins into the same machine.

How does the number of ways alter if the machine will also take quarters?

## Analyzing Students' Written Work

Even though only limited information about the problem-solving process can be gained from students' written work, some conclusions can be drawn by analyzing this work using Polya's model. One cannot evaluate the students' performance in each phase of this model. For example, evidence of the looking-back phase can seldom be found in the students' written work. If checking occurs, it is usually done mentally. Examples follow that illustrate (1) instances in which information about the student's problem-solving processes can be obtained and (2) instances in which little can be concluded from the written work. Four examples of children's written work on the following problem are given.

> At Tom's school there are 6 basketball teams. They want to plan an after-school tournament so that each team will play every other team once. How many games must be played?

Figure 1 shows that an appropriate plan has been exhibited and carried out to a correct solution. One can assume that Dina has also understood the problem adequately and has performed well in the first three aspects of problem solving.

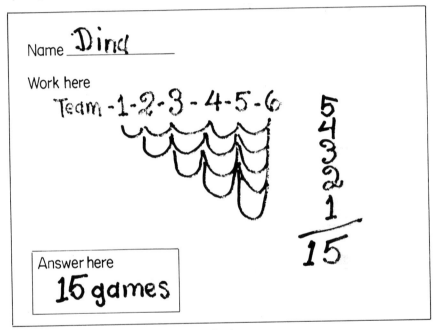

Fig. 1

If an appropriate plan is exhibited but not carried out correctly, few conclusions can be drawn about the student's understanding of the problem. The example in figure 2 illustrates this point. Was the strategy on the left of Mike's paper abandoned because it was determined inappropriate or inefficient, or because Mike did not know how to continue using it? Were the last three entries in the list at the right included because he ignored one of the conditions, or was this merely a careless mistake?

If an inappropriate plan is exhibited, any conclusions made about the understanding of the problem would usually be mere speculation. Consider the example in figure 3. Although this solution does not satisfy the condition that each team will play every other team once, Brooks's drawing indicates that possibly he does understand the problem. The particular diagram strategy that he used might have obscured the proper solution. At least we can conclude that the child has not devised an appropriate plan.

Name __MIKE__

Work here

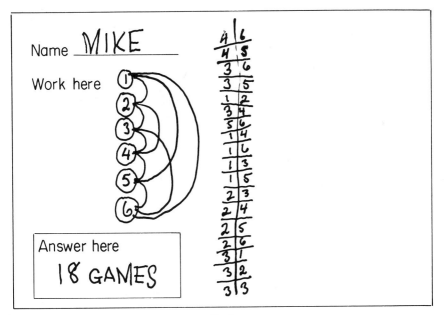

Answer here

18 GAMES

Fig. 2

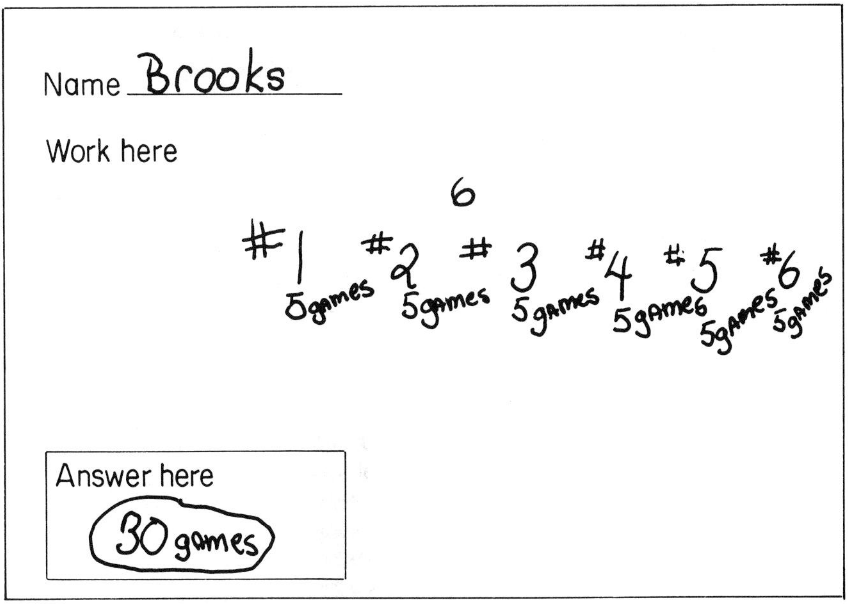

---

Name __Brooks__

Work here

6

#1          #2          #3          #4      #5          #6
5games    5games    5games    5games6  5games   5games

Answer here

30 games

Fig. 3

Only when it is clear that irrelevant or incorrect information has been used can we decide that an adequate understanding of the problem is lacking (see fig. 4).

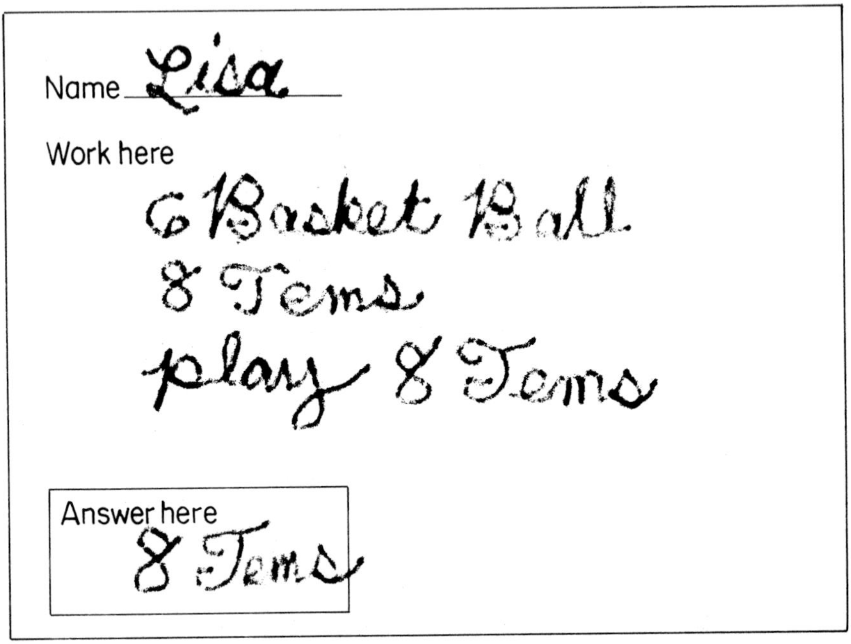

Name __Lisa__

Work here

6 Basket Ball.
8 Tems
play 8 Tems

Answer here
8 Tems

Fig. 4

It might be argued that information gained by analyzing a child's written work is primarily negative; one may be able to discover what the child cannot do but may not be able to determine what the child can do. For example, if the child has an inadequate understanding of a problem, he or she will not be able to devise an appropriate plan or carry out this plan. However, information concerning what the child cannot do can be quite useful in helping the teacher adjust instruction to meet the needs of the students.

# 12

# Problem Solving Using the Calculator

Joan Duea
George Immerzeel
Earl Ockenga
John Tarr

THE sixth-grade class had just spent twenty minutes solving story problems similar to those found in textbooks. The teacher collected the papers and was surprised to find that most students had completed more than twenty-six problems. Even more surprising, more than 90 percent of the answers were correct. Why were the students so successful in solving these problems? What had been changed? *Each student was using a calculator!*

What would happen if you tried this activity in your classroom? Reproduce thirty to forty of the usual story problems for your class. Give half your students the task of solving the problems with paper and pencil. Give the other students the same problems but allow them to use a calculator. Which group would solve more problems? Which group would have more correct answers? Which group would have more problem-solving experience? Since problem-solving skill is directly related to the number of problems solved correctly, the calculator is a significant asset.

## Advantages in Using the Calculator

### Everyone can compute

Every student can add, subtract, multiply, and divide when using a calculator. Computation difficulties inherent in paper-and-pencil work are alleviated, and the students can focus on the problem-solving process. For example, try the problem in figure 1. Time yourself as you compute with paper and pencil, and then ask a friend to try it with a calculator.

Computation does not stand in the way of solving the problem when one uses a calculator. The calculator puts the emphasis on ''what to do'' rather than on ''how to do it.'' And most important, *everyone* can do it!

The product of these two facing pages is $40 \times 41$, or 1640. Where would you open the book so the product of the two facing page numbers is 12 656?

Fig. 1

### Guess and test

Guess and test is a viable approach to solving many problems with a calculator. Students are willing to make an initial guess and reflect on the outcome when they know they can push the clear key and make a better guess.

Being willing to make a guess is a useful first step in solving both problems shown in figure 2. Furthermore, students will find that the more they use the guess-and-test approach, the better guessers they become.

Which number would you leave out so these sums are correct?

1. $42 + 65 + 18 = 107$
2. $38 + 52 + 46 = 84$
3. $53 + 47 + 38 = 85$

Which of the pairs of numbers that have a sum of 20 have the largest product?

Fig. 2

### Calculator codes

The calculator provides a new way for students to demonstrate a method of solution. A calculator code shows the sequences of keys pressed to yield the answer. A calculator code, like an equation, provides a record of the problem solver's thought processes. Furthermore, once the student develops and follows the code, the answer is displayed on the calculator.

The code that the problem solver uses for one problem will solve similar problems with only slight modifications. Figures 3, 4, and 5 use different codes. Each code could be used to solve many problems.

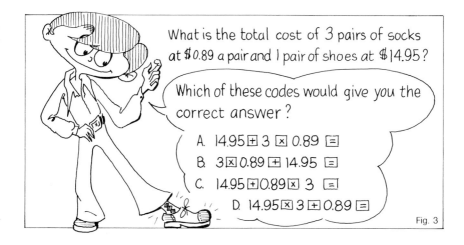

What is the total cost of 3 pairs of socks at $0.89 a pair and 1 pair of shoes at $14.95?

Which of these codes would give you the correct answer?

A. 14.95 ⊞ 3 ⊠ 0.89 ⊟

B. 3 ⊠ 0.89 ⊞ 14.95 ⊟

C. 14.95 ⊞ 0.89 ⊠ 3 ⊟

D. 14.95 ⊠ 3 ⊞ 0.89 ⊟

Fig. 3

How many seconds are there in the month of October?

OCTOBER

| SUN | MON | TUE | WED | THU | FRI | SAT |
|-----|-----|-----|-----|-----|-----|-----|
|     |     |     | 1   | 2   | 3   | 4   | 5 |
| 6   | 7   | 8   | 9   | 10  | 11  | 12  |
| 13  | 14  | 15  | 16  | 17  | 18  | 19  |
| 20  | 21  | 22  | 23  | 24  | 25  | 26  |
| 27  | 28  | 29  | 30  | 31  |     |     |

My code is _____.

My answer is _____.

Fig. 4

Write a calculator code that will give the ▨ area. _____.

Fig. 5

### Problems involving proportion

Years ago the principal method for solving proportion problems was called "the rule of three." To solve the proportion $\frac{16}{87} = \frac{x}{128}$, the student was taught to multiply $16 \times 128$ and divide the result by 87. Try using this rule and a calculator to solve the problems in figures 6 and 7.

These records all turn 45 times each minute.

Find out how long these records play.

| Turns   | 45 | 90 | 180 | 135 | 225 | 315 |
|---------|----|----|-----|-----|-----|-----|
| Minutes | 1  | 2  |     |     |     |     |

Fig. 6

A roller coaster takes a group of 24 people every 5 minutes. How long will you have to wait in line if there are 72 people in front of you?

Fig. 7

With a calculator this rule is not only efficient but also workable for all direct-proportion problems and is easy to remember. Many old procedures take on new significance when the calculator becomes a problem-solving tool.

### Real-world problems usable

Do you remember how you felt as a student when the problems you were asked to solve seemed artificial? Because of the availability of calculators, it is no longer necessary to contrive problems so that the computation fits the students' stage of development. The numbers can be real! The problem in figure 8 is an example.

Fourteen-year-old Toni Green is a pull-tab scavenger. Her collection now totals 21 000.

When clipped together like this, how long a chain would Toni's 21 000 pull-tabs make?

Fig. 8

Third-grade students studying area can solve real problems in their environment (see fig. 9) if they have a calculator. The concept of area becomes more real to them when it is not restricted by the computation involved.

Measure your classroom door. Find its area.

Fig. 9

Large numbers have always intrigued students. Students who have calculators are especially interested in solving problems that involve large numbers, and their confidence in using numbers increases. Problems like the one in figure 10 can provide practice in calculation with large numbers.

If you had $1 000 000
and you gave away $50
every hour, how many years
would it take to give away
all your money?

Fig. 10

The calculator allows students to confront consumer problems that occur in the real world. Two such problems are presented in figure 11. Students can collect actual prices from a local supermarket and use this information to find the unit cost of items, determine the better buy, and so on.

What is the unit cost for a jar of peanut butter that costs $1.09 for 360 grams? At this rate, what should a jar that contains 510 grams cost?

A liter of gasoline costs 35 cents. How much does it cost per kilometer for gasoline for a car that gets 7 kilometers to a liter?

Fig. 11

There are many advantages in using the calculator to teach problem solving, especially when numbers reflect real data.

### Increased student involvement

Students' interest in problem solving is greatly increased when they use personalized data. Students who conduct an experiment and collect their own data are intrinsically motivated to solve problems like those in figure 12.

Fig. 12

Another way to encourage student involvement is to present problem situations in which the students fill in their own data. Such problems (see fig. 13) encourage students to think about their experiences in the real world.

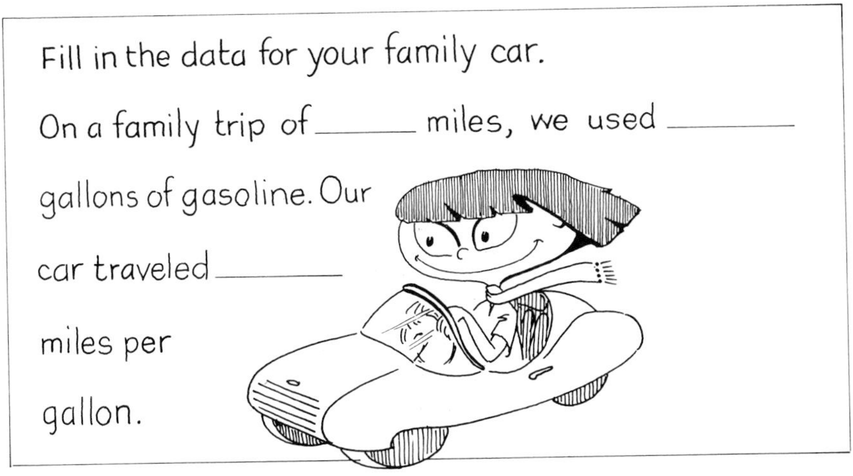

Fig. 13

The more that students feel they have an active role in creating problem situations, the more committed they are to solving the problems.

## Implementing Calculators
## in Classroom Problem Solving

Many schools purchase calculators for student use. Other schools allow their students to bring calculators from home. Whichever practice your school follows, it is most desirable to have at least one calculator for every two students.

The biggest difficulty in using calculators to solve problems is the lack of adequate instructional materials. However, present mathematics materials can be modified in a number of ways to make them more useful.

### Using the textbook

The directions that accompany most problem-solving pages instruct students to "solve these problems." Instead of giving students these directions, you can ask them to "write a calculator code to solve these problems on your calculator." This approach not only encourages the student to incorporate the calculator in the thought process but also helps the student build experiences for more difficult problems.

This technique can be extended to a number of problem-solving activities. For example, when studying circles students might try problems such as these:

1. Given the circumference of a circle, write a calculator code that will display the diameter.

2. Given the radius of a circle, write a calculator code that will display the area.

3. Given the area of a circle, write a calculator code that will display the radius.

Opportunities exist for using calculator codes at almost every level of mathematics. For example, when developing the quadratic formula, students can write a code that will solve a quadratic equation.

Textbook problems can be adapted in many ways for use with the calculator. For example, after primary students have solved the textbook problem in figure 14, have them write a similar problem with different numbers.

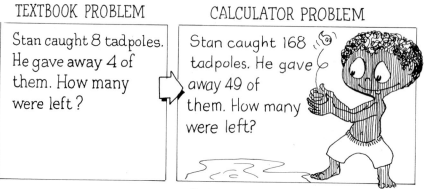

TEXTBOOK PROBLEM

Stan caught 8 tadpoles. He gave away 4 of them. How many were left?

CALCULATOR PROBLEM

Stan caught 168 tadpoles. He gave away 49 of them. How many were left?

Fig. 14

An intermediate teacher might extend a textbook problem in the manner shown in figure 15.

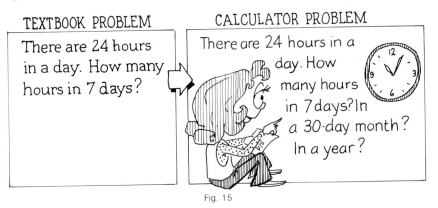

TEXTBOOK PROBLEM

There are 24 hours in a day. How many hours in 7 days?

CALCULATOR PROBLEM

There are 24 hours in a day. How many hours in 7 days? In a 30-day month? In a year?

Fig. 15

At the secondary level you can reword problems that are still in the textbook for historical reasons and thus make them more real, as in figure 16.

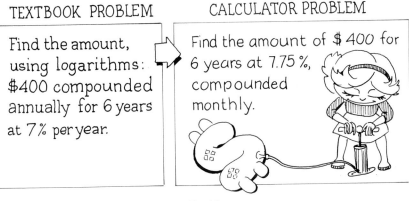

TEXTBOOK PROBLEM

Find the amount, using logarithms: $400 compounded annually for 6 years at 7% per year.

CALCULATOR PROBLEM

Find the amount of $400 for 6 years at 7.75%, compounded monthly.

Fig. 16

Students at all levels find it interesting to rewrite the problems themselves. Collecting these student problems can become an enjoyable problem-solving experience for students and teacher alike.

### A problem deck

Students do not learn much about problem solving when the problems are too easy or too difficult for them to solve successfully. Also, it is difficult to find problems that are of interest to all students. A deck of problem cards can be used to help meet students' differing abilities and interests. The deck can be organized according to the difficulty of the problems, and the problems can represent a broad spectrum of student interests.

A number of the problems in a deck can be written by the students themselves. Students often make the problems more appealing by doing their own illustrations either with original drawings or clippings from magazines. The calculator is a real help to students in both the initial writing and the solution of their problems when making the answer keys.

There are a number of ways to organize the problem deck so that it is useful in your class. If your deck is organized around the different content units in your curriculum, you may also want to structure the deck in several levels of difficulty so that students can quickly choose problems appropriate to their skills.

You may also want to organize your deck around problem-solving strategies. Build one deck of problems for which drawing a diagram or a graph is an appropriate solution and another deck for which a guess-and-test strategy is appropriate. Still another deck could contain problems for which making a table is a useful approach. When students find that a similar method works for several problems, they are more likely to remember that strategy the next time they encounter a similar problem.

We all recognize the importance of practice in teaching computation, but we forget that problem-solving skills also require practice. A deck that is organized to maintain previously taught problem-solving skills is particularly useful at the junior and senior high school levels.

There are many ways to use a problem deck. The students can form partnerships with each partnership sharing one calculator. The problem cards can then be displayed along a chalkboard tray, and each pair of students freely chooses the problems they wish to solve.

Another way to use a problem deck is to set up a problem-solving contest. Points can be assigned to the problem cards according to the level of difficulty—one point for the easiest cards and five points for the most difficult ones, for example. Students choose their problems and determine their score based on correct answers. Some students may choose to do many easy problems; others may select more difficult problems, thus scoring more points on each problem.

A third way to use a problem deck encourages the students to check their answers carefully. As the students finish each problem, they decide how sure they are of their answer. If they are "very sure" of their answer, they can wager ten points; if they "think" their answer is correct, they can wager five points; if they are "not sure," they can wager two points. After these wagers are made, the students take the time to check their work, frequently looking for an alternative method of solving the problem to be sure their answer is correct.

### Building a problem-solving file

One cannot have too many good problems. One of the best ways to collect good problems is to start a problem file. Make a file folder for each of the units in your course of study. When you find an interesting problem, drop it in the appropriate file. Old textbooks—in fact, the older, the better—are a good source of problems for your file. After a year or two you will have a good source for upgrading the problem-solving experiences in your classroom.

# 13

# Making Problem Solving Come Alive in the Intermediate Grades

**Marilyn Jacobson**
**Frank Lester**
**Arthur Stengel**

SPECIFIC attention to developing children's problem-solving abilities should begin as early in school as possible. Our experience suggests that fourth-, fifth-, and sixth-grade students can become better problem solvers if classroom activities are built around three basic principles:

1. One of the best ways to improve problem solving is through direct, active, and continuing experiences in solving a variety of problems.

2. A direct and positive relationship exists between the interest students have in a problem and their success in attacking it.

3. Successful problem-solving instruction requires an understanding of the close relationship among four distinct but highly interactive factors: students, problems, problem-solving behavior, and classroom environment.

Our bulletin-board method of applying these principles in the classroom grew out of our work with the Indiana University Mathematical Problem Solving Project (NSF grant no. PES 74-15045) during 1975–76. In cooperation with the fourth, fifth, and sixth grades of the Monroe County Community School Corporation, Bloomington, Indiana, we observed, interviewed, and taught a large number of children, and this method was one of the results.

## The Problem-solving Bulletin Board

### The ideal
Only after we spent much time with children and teachers in the intermediate grades (4–6), working in large and small groups and with and without instructional

aids, did we begin to develop a real sense of the interaction among students, problems, problem-solving behavior, and classroom atmosphere. The result of attempting to solve our own dilemma of how to make the three principles real in the classroom was the problem-solving bulletin board and its own unique instructional format.

The bulletin board, entitled "Have You Tried This?" is divided into four parts (fig. 1):

1. *The Problem*—a statement of the problem
2. *What Others Have Tried*—samples of possible methods for solving the problem
3. *Will This Help?*—questions and ideas related to the problem
4. *What I Have Tried*—a space for students to display their own attempts at solving the problem

Fig. 1. The problem-solving bulletin board

The problem statement is always visible, but the other sections of the board are designed so that only one piece of information can be seen at a time. For example, each question for "Will This Help?" is typed on a separate sheet of paper so that in order to read more than one question it is necessary to flip to the next sheet. The lower edges of the pages are graduated, and the pages are covered.

We used the bulletin board with this three-step procedure: (1) problem introduction, (2) solution effort, and (3) problem discussion. Table 1 lists some appropriate procedures the teacher might follow during each step and some desirable responses on the part of the students.

TABLE 1
Bulletin Board Classroom Procedure

| Step | Outcome | Possible Procedures for the Teacher | Possible Responses from the Students |
|---|---|---|---|
| 1. Problem presentation | Students will understand problem statement | Put problem sections on board<br>Present problem to class<br>Facilitate discussion or question asking regarding problem comprehension<br>Organize problem-solving process<br>Explain or review board sections | Read problem on board<br>Listen to problem being read<br>Write out problem<br>Record key information contained in problem<br>Ask questions to clarify problem |
| 2. Solution effort | Students will develop and try at least one strategy for solving problem | Devote class time to work on problem<br>Facilitate student efforts<br>Encourage students to help each other and share ideas | Work alone on problem<br>Read a section on board<br>Discuss problem with peers<br>Work with peer on problem |
| 3. Problem discussion | Students will discuss their respective efforts and gain new insights into problem | Allow students to demonstrate their (different) solutions or processes<br>Aid in bringing to light generalizations inherent in work on problem | Offer to discuss own process or solution(s)<br>Listen or discuss others' process or solution(s)<br>Generalize |

Here is a sample problem to illustrate the process:

Lemon drops come in packages of 3 for 10¢. Chocolate mints cost 5¢ each. Harold bought 20 pieces of candy. How many pieces of each kind of candy could he have bought?

With this problem the "Will This Help?" section would list such questions as these (one on each page):

1. How many lemon drops can you buy for 10¢?
2. How much does 1 chocolate mint cost?
3. How many pieces of candy did Harold buy?
4. Could he buy 7 lemon drops? Why?
5. What is the largest number of lemon drops he could buy?
6. Could Harold buy 20 chocolate mints and no lemon drops?

The "What Others Have Tried" section would contain samples of other children's work illustrating some possible ways to attack the problem (see fig. 2)—again, one on each page.

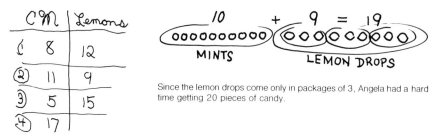

| CM | Lemons |
|----|--------|
| ① 8 | 12 |
| ② 11 | 9 |
| ③ 5 | 15 |
| ④ 17 | |

Jack began by making a list.
(CM means chocolate mints.)

Since the lemon drops come only in packages of 3, Angela had a hard time getting 20 pieces of candy.

Fig. 2. Samples of possible students' work on the lemon-drop problem

To begin the instructional procedure, the teacher—let's call her Karen—calls the class together and simply reads the problem, asking afterward if everyone understands it and if there are any questions. She answers such questions as "How many pieces did he buy?" and "How many lemon drops in a pack?" in a straightforward manner with information from the problem statement. But to questions like "Could he have bought 9 and 11?" or "Do you have to multiply?" which cannot be answered directly with information from the statement, Karen responds with such answers as "I'm not going to tell you if you're right or wrong" or "It's really your job to figure out how to solve the problem, but it sounds like you do understand what it's asking for."

After responding to these initial questions, Karen reminds the students that the problem statement, the questions (Will This Help?), and the examples of other children's efforts (What Others Have Tried) are on the bulletin board. They are free to go about working on the problem in any way they feel is best—using the chalkboard, getting into groups, or even working alone—as long as their efforts proceed in a reasonably orderly fashion.

During the next fifteen or twenty minutes the students will be working on the problem and the teacher will essentially be observing, making sure order is maintained, and dealing with questions or obstacles that arise. For example, Betty and Bob, who are working in a group, come up with different answers; Jimmy, who is working alone, wants to know if adding is the right way to get the answer. In such instances Karen tries not to be the final authority but makes an effort to direct the students to their own work, the bulletin board, or their peers to resolve the issues. Betty and Bob might be asked to show each other how they got their answers, or they might be brought together with others in their group who have still other ideas. Jimmy might be told to talk with Carla, who is busy doing computations, or Juan, who is working on a picture solution. The teacher's goal is to keep the students thinking and generating ideas, and the students' goal is to find satisfying solutions.

When the work period ends, Karen calls the class back to their seats and asks if anyone wants to share her or his work with the class. Two or three students or groups can be asked to put their work on the chalkboard or explain what they did. Using the samples on the chalkboard, Karen can discuss the different ways the problem was worked, the merits of each method, and the correct solution(s). After a suitable period of discussion, the session ends. (Karen could have had the discussion take place later in the day or postponed it till another day.)

## Observations

Our first attempts at using the problem-solving bulletin board in classrooms were not always completely successful, and certainly not every session fit the idealized process just described. But on the whole the procedure was quite successful in improving problem solving and produced some interesting observations.

*Types of problems to use*

One difficulty in providing problem-solving experiences of this type is finding problems that motivate and challenge the students. We found that the best problems were those that included one or more of these four characteristics:

1. *Problems in a context that interests the children.* Instead of asking, "How many ways can you make change for 25¢?" our problem was placed in the context of an ice-cream store:

> Art works in an ice-cream store. Ice-cream cones cost 25¢. Show the ways Art could be paid exactly 25¢.

This minor change made a major difference to the children.

2. *Problems that cannot be solved directly with a computational algorithm* (at least, not one that is known to the children). This makes them different from the typical textbook "problem," which is solved directly through the use of an algorithm that the children have been taught.

3. *Problems that can be solved by counting.* This ensures that each child can experience some measure of success. We observed that counting was the method most frequently used. For example, to solve the problem below, children typically drew a picture of a round table with X's around it (fig. 3) and then counted to get answers.

> Some children are seated at a large round table. They pass around a box of

candy containing 25 pieces. Ted takes the first piece. Each child takes one piece of candy as the box is passed around. Ted also gets the last piece of candy, and he may have more than the first and last pieces. How many children could be seated around the table?

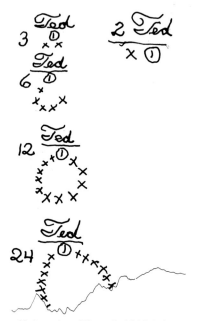

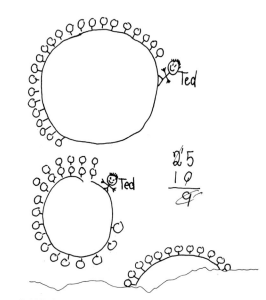

Maria made an X for each child that she thought could be seated around the table. Then she counted.

Debbie began by drawing a large round table with people around it.

Fig. 3. Samples of students' work on the box-of-candy problem

4. *Problems that (a) have multiple solutions or (b) require finding several preliminary answers to get the final solution:*

a) Show all the ways Beth could be paid her allowance of $1.60 if she received only quarters and nickels.

b) How many ways could Beth be paid her allowance of $1.60 if she received only quarters and nickels?

These problems were intriguing to the students, especially when someone would say, "That's not *all* the ways!" or "Are you sure you've found all the ways?"

At the end of this essay are some of the problems we have used successfully with fourth and fifth graders.

*The teacher's role*

We noted in table 1 that the role of the teacher is to—

1. make sure the children understand the problem;

2. create an atmosphere where working on the problem is more important than getting the right answer;

3. encourage the children to talk with each other and to work together if they wish;

4. lead a postsolution discussion in which the students share their methods.

At first some teachers were a bit uneasy regarding how much information to provide, but with a little experience they gained ocnfidence. After hearing the problem, the children tended to ask questions directed at how to solve the problem (Do we add?) or to the solution (Is it 8?). The teachers found two effective ways to deal with such questions. One was to direct the student back to the problem statements with such comments as "Is that what the problem asks you to find?" The other was to refer the questioner to other students by saying something like, "Why don't you see what Jill thinks?" As the students' successes became more evident and frequent and as the teachers polished their own techniques, the students became more independent and asked the teacher fewer questions.

During the discussion, students usually demonstrated how they had solved the problem. Other students were allowed to ask questions about the method. This sharing of solutions showed that a problem could usually be solved in a variety of ways.

At first some teachers were unsure of the extent to which they should participate and whether or not to tell the students the right answers. One teacher, Robert, had each group share its methods with the class. Even though the answers varied, Robert did not tell them which answers were correct and which were not. After a while he began to feel uneasy about this, and during the discussion of the rest of the problems he told them what the right answers were and showed them his own method of solution. Later, when we interviewed his students, they told us that the first problems they had done "didn't have answers" but all the rest of the problems did. This would indicate that it is best to end a discussion session with the students knowing which of their methods lead to a correct answer.

*Student interaction during problem solving*

It was interesting to watch children who were encouraged to work together and talk with each other in order to solve problems. After the first few problems, they developed a routine for working together. In one classroom, after the teacher had read the problem to the class and had answered their questions, the students would gather in their groups and begin working on the problem. One group would meet in a corner of the room and use the chalkboard to sketch out a solution. Another group would meet under a table. Others would work in small groups at their desks, and a few would work alone. Often the members of the group would work quietly on their own until they found a possible method of solution, and then they would share it with the group. This sharing often generated some discussion and eventually led to a solution. The students who preferred to work alone frequently checked with their classmates once they had attempted the problem on their own.

## Backward Glance

The three principles we followed and our classroom experience with the problem-solving bulletin board led to some interesting conclusions; they are stated here as propositions for your consideration:

Proposition 1. *The problem-solving success of fourth, fifth, and sixth graders can be improved by using an approach based on the three principles mentioned in the introduction.* Perhaps our most important conclusion was that problem-solving ability can be improved in almost any intermediate-grade classroom. Our procedures do not require extensive preparation or the purchase of special materials.

Proposition 2. *Problem solving is an activity best approached with a "process" perspective.* This means several things to us. Students need time to think about, and work on, problems more than they need specific instruction on how to solve them. Prior to devising the bulletin-board approach, we investigated a variety of ways to provide good problem-solving instruction ranging from attempts to teach specific strategies (e.g., looking for patterns and simplifying the problem) to attempts to develop essential problem-solving tool skills (e.g., making tables and organizing lists). None of these efforts at "teaching" children how to solve problems were effective. However, setting the proper atmosphere and supervising the flow of this peer-oriented process were successful. The teacher as "facilitator" most closely describes this notion. Answers become secondary to the process of thinking; being correct is not as important as the effort to solve a problem.

Proposition 3. *Problem-solving ability is best improved by solving many, many problems in an atmosphere conducive to exploring and sharing ideas.* Teachers should provide experiences that encourage students to try different approaches, talk with their classmates while working on problems, and discuss the relative merits of different approaches. Such an atmosphere will not only motivate students but also lead to more mature problem-solving procedures.

The process we witnessed was alive, productive, and enjoyable for both students and teachers. It evolved uniquely in each class and captured the very essence of really "doing" mathematics; yet in none of the classrooms was elaborate preparation or in-service instruction required. For these reasons we can say with some confidence, "Try it; it works!"

## More Problems

Here are some more problems! Interested? These have been used successfully in several fourth-, fifth-, and sixth-grade classrooms.

1. There were 10 handshakes at a party. You know each person shook everyone else's hand once; how many people were at the party?

2. Joe and Tim are playing a game. At the end of each game, the loser gives the winner a penny. After a while, Joe has won 3 games, and Tim has 3 more pennies than he did when he began. How many games did they play?

3. James and Judy both work part-time in a cafeteria that is open 7 days a week. James works one day and then has three days off before he works again. Judy works one day and then has four days off. James works this Monday and Judy works this Tuesday. Show the days in 3 weeks that James and Judy will work on the same day.

4. Show the ways that 15 pennies can be put in 4 piles so that each pile has a different number of pennies.

5. Show the ways that 9 marbles can be put in 5 cups so that each cup has a different number of marbles.

6. At Cathy's party there was a guessing game. Ten chips were placed in 2 cans. You win a prize if you guess the correct number of chips in each can. Show all the different guesses you could make.

7. On Monday a magic plant is 2 inches high. By Tuesday the plant has doubled its height and is 4 inches high. Each day the plant doubles its height from the day before. How high will the plant be on Friday?

8. Last night I watched a little-league baseball game. I noticed some boys and dogs playing in the grass. I heard a noise and looked to see the boys and dogs running past me. I decided to count them in a different way. I counted the legs and found there were 40. Now, what I want to know is how many boys and how many dogs ran past me?

9. Jessie collected $1.60 from each customer on her paper route. Mr. Hendron paid her with nickels, dimes, and quarters. He gave her 19 coins in all. How many of each kind of coin could he have given her?

10. Assume that a person can carry 4 days' supply of food and water for a trip across a desert that takes 6 days to cross. One person cannot make the trip alone because the food and water would be gone after 4 days. How many persons would have to start out in order for one person to get across and for the others to get back to the starting point?

# 14

# *Problem-solving Strategies in School Mathematics*

## Gary L. Musser
## J. Michael Shaughnessy

PROBLEM solving has often played a subordinate role in our heavily content-based, skill-oriented school mathematics curriculum. Even when problems do assume a central role in a course, it is seldom that the very essence of the problem-solving process—the strategies themselves—are discussed. We believe that the mathematics curriculum should become more *strategy* based and less *content* based. Students could first learn many of the problem-solving strategies for a particular content area—say, mathematics—and then later on they could see how these strategies generalize as they cut across other content areas, such as physics, biology, political science, and economics.

In the past, arithmetic has dominated much of the early school mathematics curriculum. The emphasis has been on learning and applying computational algorithms. Although much of mathematics is indeed devoted to the development of efficient algorithms to solve problems, the emphasis in this electronic age should be, not on *performing* algorithms in a rote manner, but on *developing and using* algorithms to solve problems. (Do you recall when the square root algorithm was part of the curriculum? Will our children reminisce in the same way about the long division algorithm?) Perhaps our preoccupation with computational algorithms does not serve the best interests of students. How many of them have done well through first-year algebra, only to stumble in geometry? How many have done well in most high school mathematics courses, only to have difficulty with probability? How many can differentiate and integrate in calculus without being able to use these ideas to solve problems?

If we adopt a *strategy-based* point of view, we must face some critical questions: (1) What techniques do we employ in problem solving? (2) What problem-solving strategies do we employ in school mathematics? and (3) What are some ways we can promote problem solving in the classroom?

We shall suggest in this essay some problem-solving strategies that could be incorporated in the curriculum to help nurture problem solving in the classroom. In addition, we shall suggest some ways to teach problem solving that have been used successfully by classroom teachers.

## Problem-solving Strategies

Here are some problem-solving strategies that we feel can be taught in the schools:

1. Trial and error
2. Patterns
3. Solving a simpler problem
4. Working backward
5. Simulation

Obviously, more could be included (Polya 1957; Wickelgren 1974), but let us look briefly at these five.

### Trial and error

Trial and error is perhaps the most direct method of problem solving: it simply involves applying allowable operations to the information given. There are several trial-and-error methods, two of which we shall illustrate here.

*Problem*

Find the smallest prime number greater than 840.

*Solution*

*Systematic trial and error.* If the problem is to find the smallest prime number greater than 840, begin with 841 and try each successive number, checking for prime divisors until a prime is found. Try 2, try 3, try 5, and so on. (An algorithm for checking this could be programmed on a programmable calculator.)

*Inferential trial and error.* Consider the same problem. We know that numbers greater than 840 whose units digits are 0, 2, 4, 5, 6, and 8 cannot be prime and that every third number will be divisible by 3. Thus, the first few numbers to try are 841, 847, 851, 853, 857, and 859. To see if 841 is prime we check to see if one of the primes less than or equal to 841 (other than 2, 3, and 5) divides 841. If it does, we repeat a similar procedure for the succeeding numbers until we find the smallest prime.

Inferential trial and error differs from systematic trial and error in that it takes into account relevant knowledge and uses that knowledge to narrow the search.

*Additional problems using the "trial and error" strategy*

1. Solve the following cryptarithms. In each problem, letters represent a single digit only.

a) $(HE)^2 = SHE$

b)  
```
  W R O N G
+ W R O N G
---------
R I G H T
```

2. *a)* Arrange the digits 1–9 in the triangle shown in figure 1 so that the sum of each side is 17.

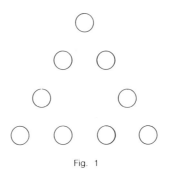

Fig. 1

*b)* Is there another "side sum" (other than 17) using the digits 1–9? What are all the possible side sums?

3. The persistence of a whole number is found as shown in figure 2. The persistence of 437 is 3, since a single digit occurs on the third step. Find numbers with persistence 4, 5, and 6.

Fig. 2

4. The following has been conjectured: Take any positive integer, $n$. If $n$ is even, divide it by 2; if $n$ is odd, multiply it by 3 and add 1. Continue to repeat this process on the result, and eventually the result will be 1. For $n = 12$,

$$12 \to 6 \to 3 \to 10 \to 5 \to 16 \to 8 \to 4 \to 2 \to 1.$$

Thus, 12 requires nine steps to arrive at 1. Find the smallest number that requires more than a hundred steps to arrive at 1. (*Note:* $2^{101}$ will arrive at 1 in 101 steps, but there is a number less than 50 that requires even more steps.)

*Solution hints*

1. *a)* We can infer that the ones digit of $E^2$ is E; so E = 0, 1, 5, or 6. But E ≠ 0 because then $(HE)^2$ would end in two 0's and H would be equal to E, which is not permitted. Also, H < 4 since SHE is a three-digit number. Now try some values.

*b)* Since there is no carry, W = 1, 2, 3, or 4. Then 2 ≤ R ≤ 9. Try W = 1 and R = 2. Then I is 4 or 5. Continue.

2. *a)* First observe that the sum of the numbers 1–9 is 45 and the sum of the three sides is 3 · 17, or 51. Thus, the sum of the corner numbers must be 6, since they were counted twice in the 3 · 17. Therefore 1, 2, and 3 must be the corner numbers. Continue.

*b)* All but two of the numbers from 17 through 23 are possible sums (Yates 1976; Carmony 1977).

3. This is a good place to combine mental arithmetic with the calculator. However, one very persistent number is 3 778 888 999.

4. Here is a systematic trial-and-error solution using a program written for the HP-33E:

| | | | | | | |
|---|---|---|---|---|---|---|
| 01 | 1 | 14 | STO 2 | 27 | f x = y |
| 02 | 0 | 15 | gFRAC | 28 | GTO 32 |
| 03 | 1 | 16 | g x = 0 | 29 | RCL 2 |
| 04 | STO 4 | 17 | GTO 25 | 30 | STO 1 |
| 05 | 1 | 18 | RCL 1 | 31 | GTO 09 |
| 06 | STO + 3 | 19 | 3 | 32 | RCL 0 |
| 07 | RCL 3 | 20 | X | 33 | f PAUSE |
| 08 | STO 1 | 21 | 1 | 34 | RCL 4 |
| 09 | 1 | 22 | + | 35 | f x ≤ y |
| 10 | STO + 0 | 23 | STO 1 | 36 | GTO 40 |
| 11 | RCL 1 | 24 | GTO 09 | 37 | 0 |
| 12 | 2 | 25 | RCL 2 | 38 | STO 0 |
| 13 | ÷ | 26 | 1 | 39 | GTO 05 |
| | | | | 40 | RCL 3 |

This program will run for about 6 1/2 minutes and the calculator will stop at the answer—27. To find the number of steps, press RCL 0—111.

## Patterns

The patterns strategy looks at selected instances of the problem. Then a solution is found by generalizing from these specific solutions.

*Problem*

Given a 1 × 1 square of toothpicks (fig 3), how many toothpicks are needed to build the 1 × 1 square into a 10 × 10 square?

*Solution*

Rather than just barging ahead and building a 10 × 10 square of toothpicks and then counting the toothpicks, suppose we solve a series of simpler problems and look for a pattern. Notice it takes 8 more toothpicks to build a 2 × 2 square (fig. 4),

Fig. 3

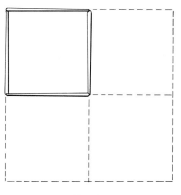

Fig. 4

12 more toothpicks to build a 3 × 3 square, 16 more for a 4 × 4 square, and so on. In this way we can solve a more general problem than the one originally given: How many toothpicks are needed to build an *n* × *n* square?

*Additional problems using the "pattern" strategy*

1. Find the sum of the first *n* odd numbers.
2. Characterize the tens digits for all powers of 3, 5, 7, 9.

*Solution hints*

1. $1 = 1$, $1 + 3 = 4 = 2^2$, $1 + 3 + 5 = 9 = 3^2$, and so on. An even nicer solution is obtained by considering the dot figures in figure 5.

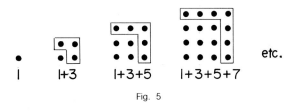

|     | I+3 | I+3+5 | I+3+5+7 |     |
| :-: | :-: | :---: | :-----: | :-: |

Fig. 5

2. First consider all powers of 5: 5, 25, 625, 3125, . . . . It appears that the tens digit will be 2 (except for $5^1$). The powers of 7 are 7, 49, 343, 2401, 16807, 117649, 823543, and so on. The pattern of the tens digits here is 0, 4, 4, 0, 0, 4, 4, and it appears that the pattern will repeat in blocks of 0, 0, 4, 4. Repeat for 3 (you may want to truncate the leading digits). Then the answer for 9 should be easy.

### Solving a simpler problem

This strategy may involve solving a "special case" of a problem or temporarily retreating from a complicated problem to a shortened version. In the latter case, the simpler-problem strategy is often accompanied by a pattern strategy. Indeed, several strategies may be required in tandem to reach a satisfactory solution.

*Problem*

How many different downward paths from A to B are in the grid in figure 6?

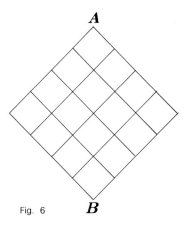

Fig. 6    **B**

*Solution*

First consider the 1 × 1 grid in figure 7. There are only two paths here. Next consider the 2 × 2 grid. In this grid, each number represents the number of paths to each respective point. Thus, there are six from A to B. Continuing in this manner will yield a total of seventy paths in the 4 × 4 grid.

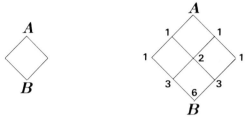

Fig. 7

*Additional problems using a "simpler problem" strategy*

1. How many squares of all sizes are in an 8 × 8 checkerboard?

2. Find all the solutions of the equation

$$\left(y^2 + \frac{1}{y^2}\right) - 18\left(y + \frac{1}{y}\right) - 17 = 0.$$

*Solution hints*

1. Complete this table and notice a pattern. (Observe how both strategies—"solve a simpler problem" and "patterns"—have been used to solve this problem.)

Size of squares

| | | 1 × 1 | 2 × 2 | 3 × 3 | ⋯ | 8 × 8 |
|---|---|---|---|---|---|---|
| Size of largest square | 1 × 1 | 1 | | | | |
| | 2 × 2 | 4 | 1 | | | |
| | 3 × 3 | 9 | 4 | 1 | | |
| | ⋮ | | | | | |
| | 8 × 8 | | | | | |

2. At first blush, this equation may appear somewhat intimidating. It is equivalent to a quartic polynomial, except at $y = 0$. However, the problem can be made simpler. Since

$$\left(y + \frac{1}{y}\right)^2 = y^2 + \frac{1}{y^2} + 2,$$

a substitution of $x = y + 1/y$ produces the quadratic $x^2 - 18x - 19 = 0$, which easily leads to the four solutions of the original equation. (*Note:* A close inspection of a problem may disclose valuable clues to its solution. The solutions to this problem come in reciprocal pairs, since if $y = b$ is a solution, then so is $y = 1/b$. One of the pairs is $y \approx 18.9472$ and $y \approx 0.0527$.)

**Working backward**

The working-backward strategy differs from the preceding ones in that it begins with the goal, or what is to be proved, rather than the given. From that point we seek a statement or series of statements that will infer the goal.

*Problem*

"Force out" is a two-person, nimlike game played as follows: A number is given, and from it the players, in turn, subtract any single-digit number. The player who is forced to obtain zero loses. Develop a system that will allow you to win, assuming you can go first each game. (Actually, you'll only be able to win nine-tenths of the time if your opponent is equally shrewd.)

*Solution*

If you can get to 1 first, you will win. To make certain that you get to 1, you must not let your opponent get to within 9 of 1, so you must get to 11 first. Continuing to work backward, you can see that you must obtain numbers whose units digit is 1 to guarantee a win. For example, if 37 is the original number, your first move would be to obtain 31. Then no matter what your opponent does you can get to 21, 11, and 1 first and force your opponent to zero.

*Additional problems using a "working backward" strategy*

1. Write a ten-digit number whose first digit tells the number of 0s in that number, the second digit tells the number of 1s in the number, . . . , and the tenth digit tells the number of 9s in the number.

2. Suppose $c$ represents the cost in dollars of manufacturing an item. A manufacturer *estimates* that the profit will be $5.00 more than twice the cost ($2c + 5$). How close to $6.00 must the cost of an item be held so that the profit is within $0.10 of $17.00?

3. One version of the game of Reverse is played as follows: The digits 1–9 are arranged in random order. The object of the game is to arrange the digits in increasing order from left to right. The only allowable operation is to "reverse the first $n$ digits from the left" when $n = 2, \ldots, 9$. For example, if R$n$ means "reverse the first $n$ digits," then 518432679 can be ordered as follows: $518432679 \xrightarrow{R6} 234815679 \xrightarrow{R4} 843215679 \xrightarrow{R8} 765123489 \xrightarrow{R6} 432156789 \xrightarrow{R4} 123456789$. Find a method (algorithm) that will make it possible for you to win. (*Hint:* You may want to consider a simpler problem first.)

4. Show that Reverse can always be won in at most $2m - 3$ moves, where $m$ is the number of digits arranged in random order. (Actually $m$ moves may be enough, but we have no proof at this writing.)

*Solution hints*

1. Consider what the solution will look like:

| 0s | 1s | 2s | ... | 7s | 8s | 9s |
|----|----|----|-----|----|----|----|
| *Digits* |  |  |  |  |  |  |

There can't be many 9s in our ten-digit number, for each time a 9 appears in a slot there must then be nine of that slot's digits in the number. For example, if 9 were the third digit *and* the fourth digit of the number, then the number must have nine 2s and nine 3s among its digits. That makes eighteen digits, which exceeds our limit of ten. Indeed, even one 9 would imply that our number repeats nine of some digit, and so could only be of the form

| 0s | 1s | 2s | 3s | ... | 9s |
|----|----|----|----|-----|----|
| $x$ | $x$ | $x$ | $x$ | ... | 1 |

for some fixed digit $x$ ($0 \leq x \leq 8$). We see that $x = 0$ won't work, since we already have a 1 (in the number-9 slot); so the number of 1s is greater than 0. If $x > 0$, then the number of 0s + 1s + . . . + 8s = 9 (digits), and so $9x = 9$, or $x = 1$.
   But

| 0s | 1s | 2s | ... | 8s | 9s |
|----|----|----|-----|----|----|
| 1 | 1 | · 1 | ... | 1 | 1 |

yields the number 1 111 111 111, which certainly doesn't have one 0, one 1, one 2, and so on, for its digits. So there cannot be any 9s in our number. In a similar way, by working backward from the number-8 slot you can show that the number of 8s is 0.

   2. The cost must be left near enough to $6.00 so that $16.90 < $2c + 5 < $17.10. Try working backward from this goal inequality so that $c$ is between two numbers.

   3. If 9 can be put in its place, it can be left fixed; hence the problem is reduced to putting 8 in its place, and so on. Working backward, we eventually get all the digits in order.

   4. The digit 9 can be put in its position in at most two moves (reverse to get 9 in the first place, then reverse all nine), and so can 8, 7, . . . , 4. The remaining three digits require three moves at most to complete the increasing sequence. Thus, ($m$ − 3) 2 + 3 = $2m − 3$ moves are sufficient.

## Simulation

   Often the solution of a problem involves setting up and carrying out an experiment, gathering data, and making a decision based on an analysis of the data. Because it may be unrealistic to actually carry out the experiment, a simulation is an appropriate and powerful problem-solving strategy.

*Problem*

   Recently the Peppy Pops Breakfast Food Company included bike-racing stickers in their boxes of Peppy Pops. How many boxes of Peppy Pops would you expect to have to buy in order to collect all five racing stickers?

*Solution*

It is not practical to solve this problem by buying samples of Peppy Pops many times until all five are collected each time. However, we could simulate the purchases by assigning to each different racing sticker a number from one to five on an ordinary die. We would ignore sixes and count the number of rolls needed so that each number, one through five, showed up. This would correspond to *one* attempt to collect all five stickers. We could then repeat the simulation many times and calculate the average number of boxes that would have to be purchased in order to collect all five stickers.

*Additional problems using a "simulation" strategy*

1. Bob had been hitting only 50 percent of his free throws for quite some time. Maurice noticed something in Bob's shooting form and made a suggestion. Bob then shot ten free throws and hit nine of them. Did the suggestion help Bob? (*Note:* Often simulation is a strategy that necessitates developing a well-posed problem from a problem situation and translating it into a mathematical environment.)

2. You and four friends are conversing at a party. Suddenly, a stranger walks up and says she is willing to bet a dollar that at least two of the people in your group have the same astrological sign. Should you take the bet?

3. Slow Sam and Fast Al are playing a finger-matching game. They throw either one or two fingers. If the number of fingers matches, Slow Sam wins; if not, Fast Al wins. Al has suggested the following payoffs for the game: if both of them throw one finger, Sam wins $10; if both throw two fingers, Sam wins $30; if the players throw a different number of fingers, Al wins $20. Al throws one finger 50 percent of the time and two fingers 50 percent of the time. Sam decides to go for the big money and shoots one finger 80 percent of the time and two fingers 20 percent of the time. Which player will come out ahead?

*Solution hints*

1. This problem situation must be developed into a well-posed problem. The problem situation is this: Did Bob hit the nine out of ten free throws just by chance, or did the suggestion help? A well-posed problem is this: What is the probability that with his *old* free-throw percentage (50 percent) Bob would hit nine out of ten? Now we can translate this well-posed problem into the mathematical environment of simulation: How often in ten tosses of a fair coin would nine heads come up? A toss simulates a free-throw attempt at the 50 percent success rate. If we toss *many* groups of ten, and see how often nine heads come up, we shall have an estimate for the probability that Bob hits nine of ten free throws just by chance.

2. If we assume that the twelve astrological signs are equally likely to occur (not quite in reality), we can simulate taking samples of five people by rolling five dodecahedral dice many times. Each die has the numbers 1 to 12 on its faces. A match of astrological signs occurs when at least two of the dice show the same number.

3. Although we can actually play this game, how could we be sure that we would succeed in making our choices on a 50-50 or 20-80 schedule? The game can be simulated using a table of random numbers. Pick a pair of two-digit numbers from

the table. Let the first number be Al's play and the second number be Sam's play. Numbers 00–49 represent one finger for Al, and numbers 50–99 represent two fingers. In a similar manner, numbers less than or equal to 19 represent one finger for Sam, and numbers greater than 19 represent two fingers. The game can be "played" many times very quickly in this manner, and winnings can be totaled at the end of a large number of plays. The next question, of course, is what is the *best* playing schedule for Slow Sam if Al maintains his 50-50 schedule? A series of such simulations can be carried out while varying Sam's playing schedule for one and two fingers. (To facilitate the simulation process, pseudorandom numbers can be generated on a programmable calculator.)

## Summary

These five problem-solving strategies—trial and error, patterns, solving a simpler problem, working backward, and simulation—can be taught in any high school mathematics program. Problems that science, government, business, and consumers face in today's world—for example, the energy crisis, rapid transportation, and inflation—require a broad range of problem-solving strategies as well as a wealth of factual information. Perhaps an emphasis on problem-solving *strategies* throughout the school mathematics program will better equip future generations for the problems they will encounter.

### REFERENCES

Carmony, Lowell A. "A Minimathematical Problem: The Magic Triangles of Yates." *Mathematics Teacher* 70 (May 1977): 410–13.

Polya, G. *How to Solve It.* 2d ed. Princeton: Princeton University Press, 1957.

Wickelgren, W. *How to Solve Problems: Elements of a Theory of Problems and Problem Solving.* San Francisco: W. H. Freeman & Co., 1974.

Yates, Daniel S. "Magic Triangles and a Teacher's Discovery." *Arithmetic Teacher* 23 (May 1976): 351–54.

*15*

# Think of a Related Problem

### Edward A. Silver
### J. Philip Smith

IN HIS writings on mathematical problem solving, Polya (1957, 1962, 1965) developed a set of heuristic suggestions that can be useful in solving a given mathematical problem. Unlike algorithms, heuristic suggestions do not guarantee a successful solution to a problem. A heuristic suggestion may be thought of as a rule of thumb, a plausible action of a general character, that may advance the course of the solution process.

Polya's suggestions are widely applicable, indicating a general course of action that might be pursued in different ways by different problem solvers and also in different ways by the same problem solver on different problems. These suggestions are based on common sense, so that they might occur naturally or easily to the problem solver. Thus, they form the basis for a reasonable approach to mathematical problem solving. In Polya's words, his heuristic suggestions "enumerate, indirectly, mental operations typically useful for the solution of problems" (1957, p. 2).

One of the heuristic suggestions that appears to have a clear relationship to classroom instruction in problem solving is "Think of a related problem which is similar to the one you are now facing." We shall focus on this single heuristic suggestion, discuss the usefulness of the advice, and offer some practical ways to assist students in using it.

## Dimensions of Problem Relatedness

As Polya has pointed out, the difficulty with the advice "Think of a related problem" is that often a great many problems are somewhat related to the one under consideration. To make effective use of Polya's heuristic suggestion, the problem solver must be able to construct or recall an *appropriate* problem related in *mathematical structure* to the problem at hand.

Even if one thinks entirely in mathematically appropriate ways, deciding on an appropriate related problem is not a simple task. Furthermore, the difficulty is compounded if one attends to the nonmathematical features present in a problem. Students often make judgments of verbal-problem relatedness on the basis of factors other than mathematical structure, as every mathematics teacher knows and recent research by Chartoff (1977) and Silver (1977) confirms.

## Nonstructural aspects

Two nonstructural aspects of a verbal problem to which many students attend are the context, or "cover story," and the question asked in the problem. Silver (1977) found that numerous students, especially the poorest problem solvers, judged problems to be mathematically related if they shared a related context—for example, problems about hens and rabbits. Thus, some students consider problems 1 and 2 to be mathematically related.

*Problem 1.* A farmer has some hens and some rabbits. Together these animals have 50 heads and 140 feet. How many hens and how many rabbits does the farmer have?

*Problem 2.* A farmer has some hens and some rabbits. He places one rabbit in a bucket and finds that it weighs a total of 4 kilograms. He removes the rabbit and places a hen in the same bucket. He finds that it now weighs a total of 5 kilograms. If the rabbit and hen together weigh 3 kilograms, how much does the bucket weigh when empty?

Another group of students, not necessarily poor problem solvers, often appeared to judge mathematical relatedness on the basis of similar question form. Some of these students would consider problem 2 to be mathematically related to problem 3, reasoning that both are "how much" problems.

*Problem 3.* In Mr. Flank's butcher shop, one pound of steak costs twice as much as one pound of bacon, and one pound of bacon costs twice as much as one pound of hot dogs. If the total cost of one pound of each is $8.75, how much does one pound of hot dogs cost?

Imagine the reactions of such students to the suggestion, "Think of a problem related to this one: 'Farmer Grundig has some hens and rabbits. The hens cost ten times as much as the rabbits. Together the animals cost $110. How much did the hens cost?' " Would the students search their memories for another "hens and rabbits" problem? Or for another "how much" problem? At this time there is no evidence that students perform such memory searches; in fact, one suspects that this totally inefficient activity would not occur. Nevertheless, it is clear that students who make judgments of mathematical relatedness on the basis of nonstructural features of problems are going to have great difficulty in constructing or recalling an appropriate related problem when the suggestion is made.

## Perception of structure

It is widely accepted that many students have difficulty "seeing" the structure of a problem, especially a verbal problem. Krutetskii (1976) found that exceptionally good problem solvers were capable of grasping the structure of a problem instantaneously, whereas poor problem solvers rarely did so. The work of Chartoff (1977) and Silver (1977) suggests that the capability of perceiving the structure of a problem is possessed in varying degrees by junior high school, senior high school, and college-aged subjects. These latter studies suggest that the perception of structure is not necessarily an ''all or nothing'' phenomenon. Perhaps more specific instructional attention should be given to this aspect of the problem-solving process.

It is fair to say that instruction related to the initial perception of problem structure is either implicit or nonexistent. In fact, teachers and textbooks may sometimes encourage students to focus on nonstructural aspects of a problem. Consider, for example, parts of a typical categorization scheme for verbal problems solved in first-year algebra—age problems, distance problems, and coin problems. It is true that the typical problems in each category share not only a related cover story but also a related mathematical structure, but our names for the categories and our procedures for dealing with such problems may call attention to the common contextual setting, not the common structure shared by the problems.

A tacit assumption in the widespread use of such a categorization scheme is that students will notice the structural as well as the contextual similarities among the problems in each category. To the extent that the assumption is valid, the categorizations can be useful; to the extent that the assumption is not valid, the categorizations may actually encourage inefficient processing of the information in the problem and may hinder students in making effective use of the advice, ''Think of a related problem.'' One wonders how students might react if they were given a problem that had a cover story associated with a typical age problem but had the structure associated with a typical river-current problem. How would students attempt to solve the problem? Which previously solved model problem would be recalled? Consider, for example, problem 4, a river-current problem.

*Problem 4.* Tom drives his powerboat downriver 700 kilometers. However, when he turns around and steers upriver, in 4 fewer hours he goes only 60 kilometers. If Tom took 10 hours to travel downriver, find Tom's rate of travel. Assume the river current and Tom move at a constant rate.

Contrast problem 4 with problem 5, a related problem.

*Problem 5.* If the combined ages of Sue's parents are multiplied by Sue's present age, the result is 700. However, when the difference between her father's age and her mother's age is multiplied by Sue's age 4 years ago, the result is 60. If Sue is 10 years old today, find the ages of her parents.

### Categorization

Evidence from work done on information processing by psychologists (e.g., Hinsley, Hayes, and Simon [1976]; Mandler and Johnson [1977]) shows that persons make early categorizations of mathematical problems and textual prose passages and that these categorizations greatly influence how persons proceed to solve the problem or interpret the passage. Thus, careful attention to how students initially perceive the important elements of a problem and how they organize their memories of experiences with problems may result in their more productive use of the advice, ''Think of a related problem.''

Some tasks, six of them discussed here, have been used in research studies and could easily be used by teachers to detect students' perceptions of problem relatedness.

*Task 1*. Students are given three problems and asked to choose the two most related problems:

Which 2 of the 3 problems below are most *mathematically related?*

**A.** A man wishes to drain his swimming pool. The pool holds 3600 liters of water. If he opens his 3 drainage pipes, they will empty the pool at a rate of 8 liters a minute. How long will it take to empty the pool?

**B.** It is a well-known fact that a lion can eat a sheep in 2 hours, a wolf can eat a sheep in 3 hours, and a dog can eat a sheep in 6 hours. How long will it take them to eat a sheep together?

**C.** A hotel can fill its swimming pool in 1 day by use of pipe A. If pipe B is used, then it takes 2 days. If pipe C is used, then it takes 3 days. If all three pipes are opened, how long will it take to fill the pool?

*Task 2*. Students are presented with a pair of problems and asked to rate them on a continuous similarity-dissimilarity scale:

Place an X on the line below to show how strongly the two problems are mathematically related.

**A.** Find two numbers whose sum is 11 and whose product is 24.

**B.** The sum of a husband's weight and his wife's weight is 135 kilograms. The product of their weights is 4250 kilograms. The husband weighs more than the wife. What is the weight of the husband?

| |
|---|
| not related **strongly** |
| at all **related** |

*Task 3*. Students are given a cue and asked to recall a previously solved problem and its solution. The cue can be some unusual aspect of the problem:

The teacher says, "Juanita, last week we solved a problem involving a man in an elevator. Tell me what you remember about that problem and its solution."

*Task 4*. Students are asked to judge problem relatedness between a previously solved problem and a new problem:

The teacher asks, "Is this problem like the milling-machine problem we solved yesterday? In what ways? Is it different in any important way? How?"

Students generally enjoy tasks like 1 through 4 because they differ from the usual classroom routine. The teacher who uses such tasks could benefit from a rich supply of data with which to plan instruction on problem solving.

*Task 5*. One very effective task that gives students the opportunity to focus on problem relatedness is to present a pair of problems and to discuss all the ways that the problems are similar and dissimilar. This activity can help students see that problems can be simultaneously related and unrelated, depending on the criteria for judgment. Class discussion could focus on those similarities and differences that have mathematical implications.

*Task 6.* Perhaps the most versatile task for classroom use is a sorting task that resembles a game in format. The task involves a set of problems, each on a separate card, that each student partitions into disjoint subsets of mathematically related problems. Students might be asked to justify their categorizations to the teacher or to each other in group discussion. Class discussions of individual criteria could result in a sharing of perspectives and might lead to a discussion of simultaneous similarities and differences. The sorting task could be used in a wide variety of ways according to the wishes of the teacher. The teacher gains the knowledge of how the students view problem relatedness and organize problems; the student gains insight into structural relationships among problems.

An abbreviated version of such a task is illustrated by the following small collection of problems:

**A.** The Yankees lead the league and the Red Sox are fifth, whereas the Orioles are midway between them. If the Indians are ahead of the Red Sox and the Tigers are immediately behind the Orioles, name the team that is in second place.

**B.** There are students on 3 committees at a certain school. The Social Committee has 21 student members, the Involvement Committee has 26, and the Service Committee has 29. There are 14 students on both the Social and Involvement Committees, 12 on both Service and Involvement, and 15 on both Social and Service. No one is on all 3 committees. How many different students are on these committees?

**C.** A pile of different-colored books sits on a library shelf. The green one is directly under the yellow one and is above the blue one. The red book is above the brown one but not next to it. The brown book is directly under the green book. Of these five books, which one is on top?

**D.** Two students each solve a certain problem. They keep a record of the time each took to solve it. The sum of the two times was 15 and the product was 36. Calculate the two individual times.

**E.** On the Yankees team 14 players each own a Spalding glove, 15 own a glove made by Rawling, and 11 own another brand. Some players own more than one glove—6 own both a Spalding and a Rawling, 4 own both a Rawling and the other brand, 2 own both a Spalding and the other brand. If no player owns 3 gloves, calculate the number of players on the team at this time.

In terms of mathematical structure the problems would probably be grouped A-C, B-E, D. Students who focus solely on context could be expected to categorize the problems as A-E, B-D, C. Those who rely on the format of the question would be likely to see A, B, and C as forming individual categories and D-E as related.

## Difficulties in Recalling Related Problems

All the previous tasks provide the teacher with knowledge about the students' thought processes. Nevertheless, the best way to gain insight into how students think is to work individually with students as they solve problems, whenever time permits. Let us discuss one recent situation and the lessons it offers.

### Nonrecognition of key elements

A first-year algebra student was solving the following problem:

The numbers below are called triangular numbers. What is the 50th triangular number?

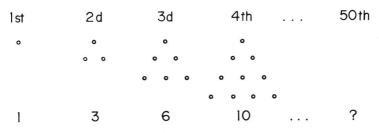

The student began by constructing the next triangular number

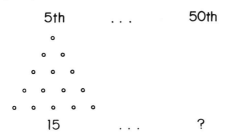

and then said, in effect, "I'll bet the answer is 150 because if you multiply 5 by 10 you get 50; so if you multiply 15 by 10 you'll get the answer 150."

With help from the teacher the student saw the lack of justification for her procedure. She then spent fifteen minutes trying to discover the fiftieth triangular number. The student recognized that the answer could be obtained with a little patience and a calculator but agreed that her goal was to find a shortcut that would avoid such an approach.

Finally, the student was clearly "looping" and making no progress. A sort of cheerful despair set in, and so the teacher made a suggestion, "Think of a simpler problem."

"Oh," the student responded, "like how you get to the twenty-fifth triangular number? . . . The tenth? . . . The sixth?"

The student went on to discover the sixth, seventh, eighth, ninth, and tenth triangular numbers. She discovered a pattern and thus solved the problem. The student and the teacher discussed the problem and its solution at the conclusion of the episode. The following exchange occurred:

*Teacher:* What do you think about the advice, "Think of a simpler question"?

*Student:* Well, even if I . . . I wouldn't have even . . . I don't know. . . . That really helped, but I think I would have eventually got to this anyway, even without that. . . . I don't know though. . . . No, I wouldn't have.

### Faulty memory

Later, thinking about this conversation, the teacher wondered, "Did the student really appreciate that heuristic advice?" The guess was no. When the student and teacher next met two days later, the problem was this:

Note the pattern below. What is the sum of the first 30 odd numbers?

$$\frac{1}{1 = 1} \quad \frac{2}{1 + 3 = 4} \quad \frac{3}{1 + 3 + 5 = 9} \cdots \frac{30}{?}$$

The student began by saying, "I'll bet the answer is 90 because if you multiply 3 by 10 you get 30; so if you multiply 9 by 10 you'll get the answer 90." The student used exactly the same invalid reasoning that she had applied to the previous problem.

Failure to recognize or remember key elements in the solution process is fairly common among students. Since this sort of behavior occurs in one-on-one sessions whose sole purpose is the study of problems and solutions, well-meant advice in the classroom situation is often likely to be consigned to some mental equivalent of the circular file.

### View of utility

If a student is to make use of a heuristic learned in connection with one problem on a second problem, the student must not only understand what the heuristic really says but also, as a minimum, believe that the heuristic can really help. Only then can we hope to get students to use heuristics when appropriate. It is unlikely to occur to a student to use a particular heuristic if the student has seen it in operation but has never been convinced of its utility. Many students are frequently so amazed or relieved or proud to have solved a problem, often with some rather crucial assistance from the teacher, that they seem unable to focus seriously on the crucial role of that assistance.

### Remedies

What are some remedies for this difficulty in recalling related problems? To make effective use of the heuristic advice, "Think of a related problem," the following conditions are necessary:

1. Students must recognize that the current problem is related to some previously encountered problem.
2. Students must be able to retrieve, from memory, appropriate information about the solution of the related problem.
3. Students must be able to translate the information into the setting and operation of the present problem.

Students therefore need experiences that will strengthen their abilities to perform the tasks listed above. They need to see repeated examples in which related problems are useful so that they will begin to think along those lines. They need training in discerning relevant relationships among problems and assistance in organizing information for later recall. Students also need training in methods of using information from previous problems.

Many teachers seem to assume that these abilities are innate and cannot be taught. Some teachers are unaware of their own thinking processes and see problem solving as something that one either "gets" or "doesn't get"—something that is generally instantaneous at crucial points. Why waste time analyzing something that is basically a one-step process, they argue.

Other teachers are rooted in a traditional approach to teaching that declares that a combination of the mathematical topic at hand, student characteristics, and teacher behavior should determine what occurs when studying problem solving. To teach problem solving, one should begin with the problems. When the problems to be solved are viewed as the most important determinant of what occurs in the classroom, then the possibility of change is greatly increased.

### Sequencing

It is necessary to carefully develop sequences of related problems if we wish to teach students about problem relatedness. The chances of influencing behavior are significantly enhanced by building instruction around problems rather than vice versa.

Students are certainly not unaware of previously solved problems. Suppes, Loftus, and Jerman (1969) found that the nature of the previous problem was the single most important variable influencing problem difficulty for students. As might be expected, problems following similar problems were easier to solve than those following dissimilar problems. Clearly, one should not expect students to transfer their learning to problems that are quite unlike those used during instruction. Many teachers, however, feel intuitively that students who first solve one problem should generally be able to solve a similar but not identical problem later. In practice, students often are not able to do so.

### Verbalization and reflection

Despite the evidence that students are influenced by previous problem-solving experience, it is far from clear how a teacher should proceed to maximize student learning or even why such learning occurs. Buswell and Kersh (1956), for example, found that students did not solve a problem in which they were asked to determine the number of x marks arranged in a triangular pattern where each row had an odd number of x's, even after solving problems like "Find the sum of $3 + 5 + 7 + \cdots + 15 + 17$." There is some indication that a reflective, thoughtful atmosphere and a discussion of the steps leading to a problem's solution are ways to help students gain from their problem-solving experiences. In studying different versions of the Tower of Hanoi problem, Gagné and Smith (1962) found that students who were required to provide a reason after each move (a process that became tedious after twenty or thirty moves!) on practice problems significantly outperformed other subjects when a related but not identical test problem was given. (The result held even though a number of the "reasons" advanced lacked insight—e.g., "only positive move," "just to try it.")

Why has research tended to indicate that such verbalization and reflection helps students become better problem solvers? The Soviet psychologist Gurova has suggested that "one necessary factor in the ability to solve a relatively complex problem requiring logical reasoning is an awareness of one's own mental operations in the problem-solving process" (1969, p. 97). One American mathematics educator even suggests defining a "good" problem solver as one who carries on an inner dialogue with himself or herself while solving a problem (Krulik 1979). Certainly the fact that an awareness of one's own thinking processes has been shown to increase problem-solving performance (Gurova 1969; Landa 1974) offers additional hope to mathematics teachers.

## A Further Look at the Advice

It might be argued that all mathematical problem-solving activity is based—often unconsciously—on a search for an appropriate related problem. Yet merely thinking of related problems is clearly insufficient for good problem-solving performance. Consider the problem "Solve for $x$ if $x^4 + 36 = 13x^2$" and the responses of three high school algebra students to the advice, "Think of a related problem."

*Student A:* How about $x^2 + 36 = 13x$? I could solve that problem as follows:

$$x^2 - 13x + 36 = 0$$
$$(x - 9)(x - 4) = 0$$
$$\therefore x = 9 \text{ or } x = 4$$

Now I could use the same method to attack $x^4 + 36 = 13x^2$:

$$x^4 - 13x + 36 = 0$$
$$(x^2 - 9)(x^2 - 4) = 0$$
$$\therefore x^2 = 9 \text{ or } x^2 = 4$$
$$\therefore x = \pm 3 \text{ or } x = \pm 2$$

*Student B:* The equation $x^2 + 36 = 13x$ is related to mine. I can solve it by using the quadratic formula:

$$x = \frac{13 \pm \sqrt{169 - 144}}{2} = \frac{13 \pm \sqrt{25}}{2} = \frac{13 \pm 5}{2}$$
$$\therefore x = 9 \text{ or } 4$$

Now I can let $x^2 = x$ so that $x^2 = 9$ or 4 and $x = \pm 3$ or $x = \pm 2$.

*Student C:* The equation $x^4 - 13x^2 + 36 = 0$ is related to the original equation, but I can't solve this one, either. How about $x^3 - 13x^2 + 36 = 0$? . . . No, I don't see any way that this can help, either. I could solve $x^2 - 13x^2 + 36 = 0$ as follows:

$$x^2 - 13x^2 + 36 = 0$$
$$-12x^2 + 36 = 0$$
$$x^2 = 3$$
$$\therefore x = \pm\sqrt{3}$$

But I don't know how to do this with an $x^4$ term and an $x^2$ term in the equation. I guess I'm really stuck! *[A suggestion is made to "try drawing a graph or a picture."]* Oh, maybe that could help. I'll make a table of values. . . . Ah, these are the solutions, $x = \pm 2$ or $x = \pm 3$."

Student A made effective use of the heuristic advice by considering an appropriate related problem, solving the related problem, and applying the *method* of the related problem to the original problem. Student B also made successful use of the advice, but this student applied the *result* rather than the method of the related problem. Student C was unable to make successful use of the advice, although the student did attempt to consider mathematically related problems. This activity alone, however, was insufficient to produce a correct solution.

Sometimes a student will be unable to solve the problem even when given the related problem. The student may be unable to solve the new problem or to relate the result to the original problem. The latter is often due to a failure to consider Polya's heuristic questions related to understanding the problem: What is the unknown? What are the data? What is the condition? Are the data sufficient to determine the unknown? There is no substitute for careful attention to these aspects of the problem.

### Generality and conscious use

The power of the heuristic advice, "Think of a related problem," lies in its general applicability across a wide range of problem-solving tasks, both mathematical and nonmathematical, and in its capacity for making us conscious of a potentially powerful tactic to use in solving problems of significance. Instruction designed to encourage students to use such heuristic advice must therefore focus on these two aspects.

Emphasizing the generality of the advice is easily accomplished if one is willing to take a fresh look at the large number of times that the advice is used in solving mathematical problems. The teacher who calls attention to his or her own use of the advice is providing an accurate model of a heuristic problem solver for students. A carefully planned sequence of problems presented over six days, six weeks, or six months can impress on students the general applicability of the heuristic advice proposed by the teacher.

The second aspect for emphasis—conscious use of the heuristic advice—is somewhat more difficult to actualize. Evidence from clinical interviews and classroom practice suggests, as mentioned earlier, that students are generally unaware of having used the methods or results of previous problems. It is a crucial task for the teacher to help students to realize how often they make use of information from related problems and to see the reasonableness of thinking of related problems when they reach an impasse in a problem-solving experience. We conclude with a few additional suggestions from research that might contribute to increased success in helping students to recall and apply the heuristic advice.

### Application to simpler problems

In a research study involving nonmathematical problems (Reed, Ernst, and Banerji 1974), students were unable to transfer information between similar problem situations unless the second problem was easier than the first. Instruction in problem solving often proceeds from a teacher-led demonstration of the solution of one or two relatively simple problems to a set of problems, some of which are more difficult, to be solved by the student. To encourage students to see the usefulness of the advice, "Think of a related problem," it might be advisable for the teacher to model the solution of a difficult problem and to suggest the application of the heuristic advice to the solution of simpler problems. After students have become convinced of the value of the advice, there will be sufficient time to progress to more difficult problems.

### Repeated suggestion

In addition to the requirement that the sequence of problems proceed from difficult to simple, Reed, Ernst, and Banerji also reported that transfer occurred only when subjects were specifically told that the problems were similar. This result suggests the advisability of telling students to look for, and think about, the possibility of an appropriate related problem. In our problem solving, we are likely to engage in such thinking; therefore, we may assume that our students do this naturally. The assumption may be *logically* sound, but the evidence from research and practical experience suggests that the assumption may be *psychologically* unsound. Careful consideration of our own use of heuristic advice, a reasonable set of problems designed to illustrate the usefulness of the advice, and the repeated suggestion by word and example that successful problem solvers use such advice may all combine to help our students make effective use of the advice, "Think of a related problem."

### REFERENCES

Buswell, G. T., and B. Y. Kersh. *Patterns of Thinking in Solving Problems.* University of California Publications in Education, vol. 12, no. 2. Berkeley: University of California Press, 1956.

Chartoff, B. T. "An Exploratory Investigation Utilizing a Multidimensional Scaling Procedure to Discover Classification Criteria for Algebra Word Problems Used by Students in Grades 7–13." (Doctoral dissertation, Northwestern University, 1976.) *Dissertation Abstracts International* 37 (1977): 7006A. (University Microfilms no. 77-10,012)

Gagné, R. M., and E. C. Smith, Jr. "A Study of the Effects of Verbalization on Problem Solving." *Journal of Experimental Psychology* 63 (1962): 12–18.

Gurova, L. L. "Schoolchildren's Awareness of Their Own Mental Operations in Solving Arithmetic Problems." In *Problem Solving in Arithmetic and Algebra,* edited by Jeremy Kilpatrick and Izaak Wirszup, pp. 97–102. Soviet Studies in the Psychology of Learning and Teaching Mathematics, vol. 3. Chicago: University of Chicago Press, 1969. (Available from the National Council of Teachers of Mathematics, 1906 Association Dr., Reston, VA 22091.)

Hinsley, D. A., J. R. Hayes, and H. A. Simon. "From Words to Equations: Meaning and Representation in Algebra Word Problems." (CIP Working Paper no. 331.) Unpublished manuscript, Carnegie-Mellon University, 1976.

Krulik, Stephen. "Problem Solving—You *Can* Teach It." *Frostburg State College Journal of Mathematics Education,* no. 16, Fall 1979. (Proceedings of the Frostburg State College Mathematics Symposium, Frostburg, Maryland, April 1978.)

Krutetskii, V. A. *The Psychology of Mathematical Abilities in Schoolchildren.* Edited by Jeremy Kilpatrick and Izaak Wirszup and translated by Joan Teller. Chicago: University of Chicago Press, 1976.

Landa, L. N. *Algorithmization in Learning and Instruction.* Englewood Cliffs, N.J.: Educational Technology Publications, 1974.

Mandler, J. M., and N. S. Johnson. "Remembrance of Things Parsed: Story Structure and Recall. *Cognitive Psychology* 9 (1977): 111–51.

Polya, G. *How to Solve It.* 2d ed. Princeton, N.J.: Princeton University Press, 1957.

———. *Mathematical Discovery.* Vols. 1 and 2. New York: John Wiley & Sons, 1962, 1965.

Reed, S. K., G. W. Ernst, and R. Banerji. "The Role of Analogy in Transfer between Similar Problem States." *Cognitive Psychology* 6 (1974): 436–50.

Silver, E. A. "Student Perceptions of Relatedness among Mathematical Word Problems." Unpublished doctoral dissertation, Columbia University, 1977.

Suppes, Patrick, E. Loftus, and Max Jerman. "Problem-solving on a Computer-based Teletype." *Educational Studies in Mathematics* 2 (1969):1–15.

# 16

# *Risking the Journey into Problem Solving*

## Peggy A. House

**M**ATHEMATICS pupils behave just about as we expect: when we expect them to be mathematically incapable, they rarely let us down; when we expect them to achieve, they generally come through. But expecting pupils to achieve involves taking risks. This is especially true when it comes to problem solving.

Having pupils engage in open-ended problem solving is one way of giving them valuable experiences in the processes of mathematics, but it means we must risk some of the security of knowing in advance just how the lesson will proceed or exactly what the answers will be. Students are accustomed to encountering mathematics in its finished form. Problem solving is mathematics in the making. And until pupils and teachers both create considerable mathematics, both will be threatened. Here are some practical hints to help teachers become more spontaneous problem solvers and leaders of problem solving. Each is illustrated with a concrete example taken from actual experiences of classroom teachers.

Before we proceed, however, one danger must be carefully noted and consciously resisted. This is the tendency to deny ourselves the freedom to investigate or to experience problem solving because "problem solving takes too much time." (Equivalent statements include "We won't cover the syllabus," "We have to have time for the basics," "We have to keep pace with the other classes," etc.). Let us assume that we want to make problem solving an important dimension of our classes, and so we shall temporarily suspend any "yes buts" or "what ifs" that threaten to truncate this process.

### How else could I ask that question?

We observed that students usually confront mathematics in its finished form rather than mathematics in the making. Polya has differentiated between these two by noting that finished mathematics consists of proofs whereas mathematics in the making consists of guesses. We might paraphrase this to suggest that mathematics in the making consists of questions. And the way we ask a question frequently determines whether we shall be embarking on a problem-solving adventure or on an exercise. Hence we must continually think, How else could I have asked that question?

An example will illustrate the difference. While Ellen, a teacher, was preparing an assignment, she encountered the following question in the text: "Try to construct squares with areas 1, 2, 3, 4, 5, 6, 7, 8, 9, and 10 on your geoboard. Exactly three of these squares cannot be constructed on the geoboard. Which ones are they?"

The manner in which the question and hint were presented aborted the potential problem-solving process. The teacher herself approached the task as most students would: she listed the numbers 1 through 10 and crossed off those that corresponded to the area of squares she knew she could make—1, 4, 9, 2, 8, and so on—until three numbers remained. Then she went on to the next exercise. The question was answered. The book, after all, had said there would be three impossible ones.

But in this case Ellen went back and asked, "How could I ask that question differently?" And when she asked it differently, she launched a problem-solving adventure that engaged her for months.

First she asked, "What squares can I make?" She already knew of seven. Were there more? Then she asked, "Are squares of areas 3, 6, and 7 really impossible?" She posed this new task to her students: Either demonstrate that you can make the square or prove that you cannot. The next day the students returned with at least three different proofs—one based on the Pythagorean theorem (use the theorem to determine the lengths of the fourteen line segments that can be constructed and show that none equal $\sqrt{3}$, $\sqrt{6}$, or $\sqrt{7}$); one based on Pick's rule (use the formula Area $= \frac{1}{2}(B - 2) + 1$ to determine what combinations of boundary ($B$) and interior ($I$) pegs would be necessary and then show that none of these can be constructed); and one based on construction (using any peg as center, construct circles of radii $\sqrt{2}$ and 2; since there are no pegs between the two circles, there can be no side of length $\sqrt{3}$).

Next she asked, "What other squares can I make on my geoboard? What squares could I make if my geoboard had ten pegs on a side instead of five?" The geoboards in the classroom proved to be too small to answer the question fully, but with the aid of dot paper she could investigate patterns of possible and impossible squares on any $n \times n$ geoboard. This led naturally to asking *how many squares* and *how many noncongruent squares* could be made on the $n \times n$ board, and it involved an awareness that these are two different questions with different answers. It also opened a class discussion on number patterns.

From these considerations of $n \times n$ geoboards, Ellen took another step. Suppose the geoboard was unbounded: what would be the possible and impossible squares? By afternoon she had found several families of possible squares, each having sides of length

$$k(N^2 + 1)^{1/2}$$

and the area of

$$k^2(N^2 + 1),$$

where $k$ and $N$ are positive integers. She also had noted that for fixed $N$, the value $k = 1$ gave the unit square for that family, and she found that each unit square contained in its interior an array of pegs forming a square of side $(N - 1)$.

Several days later it occurred to Ellen that her solution was incomplete—that the families of squares she had found thus far all belonged to a single class of squares and that there were many more such classes. Subsequent investigation of other classes of squares led her into an exploration of number patterns and into the surprising discovery of many artistic patterns and designs generated by the square. It also led her and one of her students to an exploration of a new divergent series and to a correspondence between the geoboard problem and an unrelated question about complex numbers.

Finally, the original question led the class in another direction when Ellen suggested that they remove the restriction that the squares be formed by using a single rubber band. This way the students could form squares by the intersection of parallel lines. Such squares need not have their vertices at lattice points, and consequently a whole new problem area was opened.

In short, a very closed, convergent question, when asked differently, was turned into an open-ended, divergent learning experience.

### What other problems does this suggest?

In addition to asking, "How could I ask that question differently?" we must also develop the habit of asking, "What other problems does this suggest?" Take, for example, another teacher, Norma, whose students had solved the problem of arranging the integers 1 through 6 in a triangular array

$$X$$
$$X\ X$$
$$X\ X\ X$$

so that each side of the triangle added up to 9. The students had also discovered that the numbers could be rearranged to give sums of 10, 11, or 12. In the process they had noted further that for each desired sum, $S$, they could write exactly three equations of the form

$$a + b + c = S.$$

An examination of the three equations immediately revealed which three numbers must go in the vertex positions. The rest was obvious.

But to Norma it posed a new question: What would happen with a larger triangle—one of, say, nine points?

$$X$$
$$X\ X$$
$$X\quad X$$
$$X\ X\ X\ X$$

A colleague suggested that the sum should be 23; so Norma began, as in the simpler problem, to generate equations of the form

$$a + b + c + d = 23,$$

where $a$, $b$, $c$, and $d$ were integers from 1 through 9. To her surprise, there were nine such equations, and the choice of vertex numbers was not obvious. By trial and error she found three equations that produced the desired triangle.

But what of the remaining six equations? Did they also yield solutions? And what about other sums? What besides 23 is possible?

The first question led her to discover a different triangle, which also had a sum of 23 along each side and which was formed from three different equations in the list. The remaining three equations, however, produced no solution.

The concern for other possible sums led her to a theoretical prediction that the possible sums ranged from 17 to 23. Through trial and error she found the following results:

| Sum | Number of Solutions |
|-----|---------------------|
| 17 | 2 |
| 18 | 0 |
| 19 | 4 |
| 20 | 6 |
| 21 | 4 |
| 22 | 0 |
| 23 | 2 |

This led next to the question, Why are 18 and 22 impossible sums? By means of an indirect proof, Norma established the necessary contradiction to verify their impossibility.

But the exploration had only begun. By continuing to ask, "What extensions does this result suggest?" she was able to conclude that the sum of the three vertex numbers must be divisible by 3. Further, this sum, $V$, is such that $6 \leq V \leq 24$ when the corresponding sum, $S$, on any side is $17 \leq S \leq 23$.

Next she examined the various triangles, such as the one below, by using the definitions that follow it:

$$a$$
$$b \quad c$$
$$d \quad\quad e$$
$$f \quad g \quad h \quad i$$

vertex numbers: $a$, $f$, and $i$

side numbers: pairs of nonvertex numbers on the same side ($b$ and $d$; $c$ and $e$; $g$ and $h$)

corner numbers: pairs of numbers adjacent to the same vertex number ($b$ and $c$; $d$ and $g$; $e$ and $h$)

diagonal numbers: noncorner pairs of numbers on opposite sides of the triangle which can be joined by a line parallel to the third side ($d$ and $e$; $b$ and $h$; $c$ and $g$)

She found that the two solutions for $S = 23$ could be written as follows:

$$9 \quad\quad\quad\quad\quad\quad\quad 9$$
$$5 \quad 3 \quad\quad\quad\quad\quad 2 \quad 6$$
$$1 \quad\quad 4 \quad\quad\quad\quad 4 \quad\quad 1$$
$$8 \quad 6 \quad 2 \quad 7 \quad\quad 8 \quad 3 \quad 5 \quad 7$$

She and her seventh-grade class made the following observations:

1. The vertex numbers are in arithmetic progression: 7, 8, 9.

2. The sums of pairs of vertex numbers are 15, 16, 17.

3. The pairs of side numbers sum to 6, 7, 8.

4. The pairs of corner numbers sum to 6, 7, 8.

5. The pairs of diagonal numbers sum to 5, 7, 9.

6. Each pair of side numbers has the same sum as the opposite pair of corner numbers.

7. Each pair of corner numbers sums to one less than the corresponding vertex number.

8. The vertex numbers sum to one more than the sum of any side.

9. The first triangle can be transformed into the second by interchanging the pairs of diagonal numbers.

10. The solutions for the two triangles having a sum of 17 on any side can be obtained by replacing each $n$ above with $(10 - n)$.

The last two observations led to a new discovery: The four solutions for each of the sums 19 and 21 can similarly be written in pairs where one solution for a given sum is obtained from the other by interchanging diagonal numbers, or where a solution for $S = 19$ is obtained from a solution for $S = 21$, and vice versa, by replacing corresponding entries, $n$, with $(10 - n)$. In the case of $S = 20$, four of the six solutions formed two pairs that are related by the diagonal interchange method. The relationships between the other two were more complex.

By combining all the information in a table, Norma discovered a constant, $k$, associated with each value of $S$ where

$$S = V + k;$$

$k$ was always an odd integer such that

$$-1 \leq k \leq 11.$$

Having discovered $k$, she was able to generalize a new set of patterns. The list soon included more than ten generalized patterns, two of which follow:

1. For each vertex number, $v_i$, if $S^*$ is the sum of the opposite side numbers, then

$$S^* = v_i + k.$$

2. In any triangle, the difference between any two vertex numbers is equal to the difference between the sums of the opposite pairs of side numbers.

Once again an extensive problem-solving experience emerged from an open-ended asking of the repeated question, What more?

### Am I missing the mathematics because of the arithmetic?

"My eighth graders asked the strangest question today," remarked Charles to his colleagues after school. "They wanted to know how much money I make an hour." Charles had tried to explain that he worked under a contract, not on an hourly basis, but the class had insisted that he tell them what his hourly pay was. At length he had given in and with the help of the class divided his annual salary by the product of the number of school days in a year and the number of hours in a day. Finished. Now back to the day's lesson.

Another teacher in the group pointed out that this was arithmetic rather than mathematics and suggested that they might be able to turn it into an opportunity for problem solving by generating some open-ended questions. At first the teachers were skeptical about the relationship of this situation to their mathematics classes, but they decided to brainstorm for a few minutes. They made a list of questions including the following: What besides cash is a part of one's earnings? How much do others make? How much of one's salary goes for housing, food, clothing, taxes, and so on? What is meant by *cost of living?* What contributes to it? If I got a raise of $1000 this year, how much would I actually realize? What is inflation all about? At the current inflation rate, what will it cost in ten years to buy goods valued today at $100?

Each question suggested opportunities to apply and practice the arithmetic of computation, fractions, decimals, percentage. But the teachers wanted to look beyond arithmetic. They focused on their second question: How much do others make?

First they would have to define an acceptable sample of "others." What variables should they consider? Some they suggested included (*a*) type of work (professional, skilled, semiskilled, etc.); (*b*) type of employer (government, private industry, etc.); (*c*) length of service; (*d*) age; (*e*) sex; (*f*) race (member of a minority?); (*g*) position in the hierarchical structure of the business; (*h*) geographic region of the country; and (*i*) setting (urban, small town, or rural).

The problem began to take shape, but it was too big. In an attempt to make it both manageable and relevant, they decided to investigate those careers in which their students expressed interest. All eight teachers returned to their respective classes and pursued the question with their pupils.

During the next few weeks the classes gathered data; organized the data in tables, charts, and graphs; compared salaries; computed ratios and percentages; made predictions; designed and conducted surveys; and discovered a wide range of new applications. The pupils surprised their teachers with their interest in the project; this reaction in turn enhanced the teachers' enthusiasm, and the mutual stimulation of their curiosity motivated them to continued their investigations. As a result, what emerged was not an exercise in arithmetic but a rich experience in mathematics.

### Let me do it my way!

How often have we been guilty of creating the impression that in mathematics there is one right way to solve a given type of problem? How many pupils, former pupils, and even teachers believe, for example, that there is one "correct" algorithm for division or multiplication? Our fourth consideration is a frontal attack on this misconception: encourage pupils to find multiple solutions for problems, and reward the unique or original or creative.

A favorite problem comes immediately to mind: Given square *ABCD* with *M* the midpoint of $\overline{CD}$. $\overline{BM}$ intersects $\overline{AC}$ at *E*, forming regions *P*, *Q*, *R*, and *S*. Find the ratio of the areas of *P*, *Q*, *R*, and *S* (or, alternatively stated, if the area of *ABCD* is one unit, find the areas of *P*, *Q*, *R*, and *S*.) (See fig. 1.)

Here are some approaches offered by students from seventh grade to graduate level.

*From a seventh grader:* Cut out a triangle congruent to *P* and trace that triangle in *R* "by sliding the cutout into each corner," thus "establishing" the ratio of *P* to

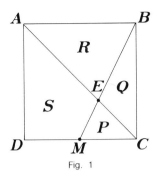

Fig. 1

$R$. (Note that with seventh graders the teacher was not immediately concerned with a rigorous proof of the similarity of $P$ and $R$; an intuitive demonstration was considered appropriate.) Cut out $ABCD$ and fold along $\overline{AC}$ to show that $P$ and $Q$ have equal altitudes. (The rest followed without difficulty.)

*From a junior high school teacher:* Let $N$ be the midpoint of $\overline{BC}$, and draw $\overline{DN}$. (See fig. 2.) From the symmetry of $\overline{DN}$ and $\overline{BM}$, show that $Q$ is divided into two triangles of equal area, one of which is congruent to $P$. Establish the ratio of $P$ to $R$ directly from a knowledge of similar triangles. The rest follows.

*From a high school senior:* Establish that the area of $R$ is four times the area of $P$ from a knowledge of similar triangles. Show that the altitudes of $P$ and $Q$ are equal, since $E$ is on the bisector of $\angle C$; hence, show that the area of $Q$ is twice the area of $P$. Draw $\overline{AM}$, dividing $S$ into regions $S_1$ and $S_2$ (see fig. 3). Show that the area of $(S_2 + P)$ equals the area of $(P + Q)$. Hence, $S_2$ and $Q$ have equal areas. Also, $S_1$ has three times the area of $P$, since the altitude of $P$ through $E$ is one-third the length of the side $\overline{AD}$. Thus, express all areas as multiples of $P$ and solve.

*From an algebra student:* Let the sides of the square be $s$. Then, for example, $P$ has base $s/2$ and height $s/3$. Describe each region similarly and solve algebraically.

Other solutions included a trigonometric analysis in which a graduate student arrived at areas of the regions by applying appropriate sine and cosine relationships. An undergraduate constructed all eight segments from the vertices to the midpoints of the opposite sides and solved the problem by determining areas of some of the

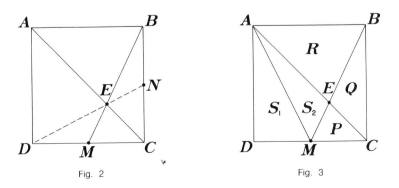

Fig. 2          Fig. 3

triangles and other polygons thus formed. Many other solutions combined elements of several of these. Each was unique to the solver.

### How far can you take that?

One characteristic of a good problem for the classroom is flexibility: the problem lends itself to many levels of solution. This makes it possible for students of differing ability and insight to experience some degree of success as each solves the problem to a different level of abstraction or generalization. A seventh-grade class demonstrated this during a lesson in which the teacher posed the following problem:

In a certain high school there were 1000 lockers and 1000 students. Each year on Homecoming Day the students lined up in alphabetical order and performed the following strange ritual: The first student opened every locker. The second student closed every second locker. The third student changed every third locker (i.e., opening the closed lockers and closing the open ones). The fourth student changed every fourth locker, and so on. After all 1000 students had passed, which lockers were open?

Immediately the class began to guess: "All of them." "None of them." "Only the first one." "All the even numbers." "All the odd numbers." Clearly, it would be necessary to find a more systematic approach. One student wanted to draw a picture. After discussing what the picture should illustrate, the class decided to make their picture as shown in figure 4.

The teacher and pupils together completed the first three rows using the overhead projector; then each pupil continued the investigation individually. All discovered that lockers 1, 4, 9, and 16 were open, and almost all recognized that between the open lockers there were first two, then four, six, eight . . . closed lockers.

Many were content to stop at this point. For them the problem was solved. But the teacher continued the challenge: What is the next open locker after #16? Will #100 be open or closed? How about #500? Is there a way to find out without drawing the solution for all 1000?

Amy offered an observation: All the lockers with prime numbers are closed. She also could explain why. Neil observed that the only students to touch a given locker were those whose number was a factor of the locker number. Further, a number would have to have an odd number of factors if that locker was to remain open.

Only a few pupils recognized that the open lockers were the perfect squares. Once they made that observation, however, they were able to determine if any given locker would be open or closed. A few pressed on to ask *why* the perfect squares had an odd number of factors. By writing the factors of 24 and 36, they saw that all factors except the square root occurred in pairs. A few were curious about why some numbers had many factors and others did not. This opened the door to investigations of the conditions under which a number would have exactly 2, 3, 4, . . . factors. The observation that every locker visited by student 4 was also visited by student 2 but that the reverse was not true led to a discussion of least common multiples and greatest common factors.

Not every pupil in the class was interested in all the generalizations and extensions, but each left the class believing that he or she had helped solve the problem satisfactorily.

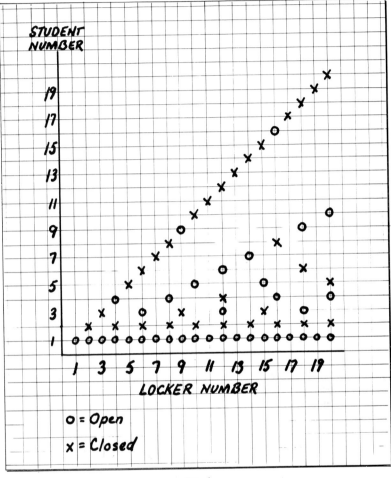

Fig. 4

### I'd rather do it myself!

One of the hazards of being an adult is the inherent desire to show children how to do things more quickly, more efficiently, more "correctly." In the mathematics class, this tendency frequently makes it difficult for teachers to withhold "telling" while patiently allowing pupils to struggle with a problem. Pupils, however, generally expect that they are to find a definite *right* answer, and they look to the teacher to provide sufficient hints to guarantee their success.

A useful method of contending with this narrowly focused search for *the* answer is to present problems that have no particular answer. An excellent opportunity lies in pattern searches. Elementary pupils enjoy looking for patterns in the 100 chart, the 10-by-10 multiplication table, or Pascal's triangle. Older pupils might generate the sequences of triangular, square, pentagonal, hexagonal, . . . numbers by arranging chips or markers as shown in figure 5:

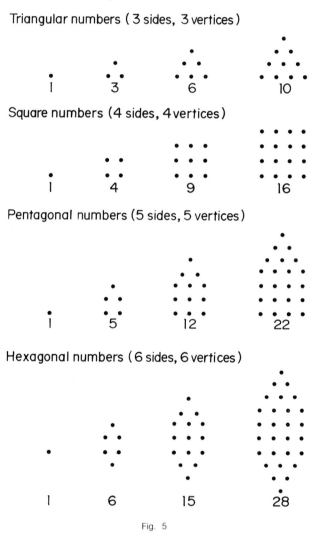

Fig. 5

These sequences are known as the figurate numbers, which we designate as $_kF_n$, where $k$ tells the number of sides or vertices and $n$ represents the position in the sequence. Students can combine these numbers in a figurate number table (table 1).

The search for patterns and relationships within the table can yield rich insights. Here are a few that pupils have found:

1. The triangular numbers differ by successive integers.
2. The square numbers differ by successive odd integers.

TABLE 1
Figurate Numbers: $_kF_n$

| k \ n | 1 | 2 | 3 | 4 | 5 | 6 | • | • | • |
|---|---|---|---|---|---|---|---|---|---|
| Triangular numbers ($k = 3$) | 1 | 3 | 6 | 10 | 15 | 21 | • | • | • |
| Square numbers ($k = 4$) | 1 | 4 | 9 | 16 | 25 | 36 | • | • | • |
| Pentagonal numbers ($k = 5$) | 1 | 5 | 12 | 22 | 35 | 51 | • | • | • |
| Hexagonal numbers ($k = 6$) | 1 | 6 | 15 | 28 | 45 | 66 | • | • | • |

3. The differences down any given column are constant, and these differences are the same as the triangular numbers.
4. Entries in the odd-numbered columns are multiples of the column number.
5. After the first row has been completed, any number can be obtained by adding the entry just above it to the triangular number in the preceding column (i.e., $_kF_n = {_{k-1}F_n} + {_3F_{n-1}}$).

Because there are no particular predetermined answers, any patterns the pupils discover are acceptable. (Some classes will choose to dedicate certain relationships to their discoverer, as in "Pat's rule.")

Teachers, too, can participate in real problem solving and discovery; this is an important corollary to the do-it-myself idea. Teachers frequently remark that one reason they do little or no problem solving in their classes is that they themselves are uncomfortable or insecure as problem solvers. Like their pupils, however, teachers become problem solvers by solving problems. A do-it-myself procedure, such as investigating numbers in an open-ended, divergent way, is one effective means of building self-confidence and enthusiasm for problem solving.

## Help me, please!

A teacher's enthusiasm for problem solving is almost always contagious. Pupils who see that their teachers are enthusiastic about solving problems that intrigue them tend to cultivate their own natural curiosity and eagerly participate in the teacher's problems. It is not necessary for the teacher to know the answer before sharing a problem with the students. Often, in fact, the problem arises unexpectedly.

Students in a college methods course were discussing classroom uses of the pocket calculator. They had inquired about the decimal expansion of 1/7, a trivial question if the calculator has an eight-digit display. Then they used their calculators to determine the complete decimal expansions of 1/17 and 1/19 and related both to their observations of the behavior of 1/7. Wondering why the instructor had skipped 1/13, one student tried to apply the same technique and found to his surprise that the period of the repeating decimal contained only six digits instead of the twelve he had expected. Although the instructor was not prepared with a ready explanation of the phenomenon, he recognized it as an opportunity for shared problem solving, and both teacher and student tackled the problem together. Others who overheard them joined in, and for several weeks students would stop in periodically to add some new insight to the shared investigation. Eventually, they pieced together a satisfactory explanation. (For a fraction $1/n$ where $n$ is prime, the number of digits in the period of the decimal expansion is the smallest positive integer, $k$, for which $10^k - 1$ is divisible by $n$.)

### Please pass the problems

Obviously not all problems emerge as spontaneously as in the example above. Most are deliberately introduced by teachers and, we hope, by pupils. This demands a problem resource, which teachers will want to begin now to build.

Collecting problems on index cards is a recommended approach because the cards can be filed according to content, useful heuristics, age/grade level, and so on. Some teachers prefer to write the problems on poster paper so they can be quickly pinned to the bulletin board as a new "problem of the week." Any system of collecting problems is a good one if it gets them into the classroom.

Many good sources of problems exist. Books of puzzles and problems are in every bookstore. (See the Bibliography in this yearbook.) The writings of George Polya and Martin Gardner are especially recommended. Newspapers frequently run puzzle columns. Airline magazines have yielded some good resources. Trading problems with other teachers is a rich source of new material. And some of the best problems will be those that students and teachers themselves create or discover in their day-to-day lives. (One student who had gotten his new license plates came to class and asked, "How many more plates can be issued this year, since the state has changed from two letters and four numerals to three letters and three numerals?")

### Conclusion

Problem solving is an integral part of the mathematics class, not an adjunct or a filler for days before vacation. By consciously engaging pupils in the process of making mathematics, we shall, we hope, help them to identify and select relevant information; search for patterns, relationships, and generalizations; formulate plans and procedures; integrate and employ previously learned concepts and skills; and extend their learning to new situations. In short, we shall help them to become makers and users, not just observers of mathematics. When we do this regularly, we find that the risks we take pay off in positive outcomes for both teachers and pupils; we discover too that not only teachers but also pupils come to expect and to find enjoyment and achievement in mathematics.

# 17

# Problem Solving through Recreational Mathematics

## Kevin Gallagher

Several highly developed mathematical disciplines began as purely recreational pursuits: combinatorics, game theory, number theory, and topology. In fact, virtually every field in mathematics has recreational aspects. Unfortunately, no well-defined set of mathematical topics is universally accepted as recreational; what is recreation for one individual is work for another. (See Trigg [1978] for an excellent discussion and reply to the question "What is recreational mathematics?") Nevertheless, many mathematical topics *are* viewed as recreational by a large number of enthusiasts. William L. Schaaf's four volumes (1970; 1970; 1973; 1978) entitled *A Bibliography of Recreational Mathematics* attest to this.

Problem solving is the one theme found common to most topics in recreational mathematics. Frequently, extensive discussions of problem-solving strategies can be found in the pages of recreational mathematics literature. The numerous books published in recreational mathematics are primarily collections of problems. A typical issue of the *Journal of Recreational Mathematics* is heavily problem oriented. Indeed, to be active in recreational mathematics is to be an active problem solver.

For the purpose of instruction in problem solving at the high school level, two topics from recreational mathematics are considered here: strategy games and mathematical puzzles. In a strategy game the focus is on finding a winning strategy. In a mathematical puzzle, the focus is on finding a solution that uses a minimum of sophisticated mathematical tools and is, at the same time, concise and easy to understand. When a student encounters a solution, there should be instant recognition that the solution is obvious and does indeed solve the problem. The difficulty with a puzzle rests, not with the solution, but rather with the solution-finding process.

(The meaning of "mathematics problem" is limited here to include only applications of mathematics to real-world problems or to supposed real-world problems. Although the definition could be generalized to include abstract mathematical problems, the typical notion of mathematical problems in high school is more closely associated with applications. This limited definition is therefore more appropriate.)

The chart in figure 1 shows the relationship between general mathematical problem solving, as defined above, and that encountered in strategy games and mathematical puzzles. A solution to a particular problem, game, or puzzle may differ somewhat from the steps as they are outlined. The steps are not necessarily discrete or sequential. With this in mind, note that they are, nevertheless, indicative of the processes involved.

An example and brief discussion of a strategy game and of a mathematical puzzle follow.

## A Strategy Game

The L-Game was invented in the mid-1960s by Edward DeBono of Cambridge University. It is very simple in design. The board consists of sixteen squares arranged in a 4 × 4 square (fig. 2). There are two L-shaped playing pieces of different colors covering four squares each, one for each player, and there are two neutral pieces of a third color covering one square each, belonging to neither player.

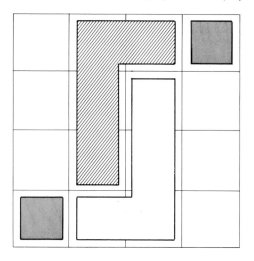

Fig. 2. The L-Game starting position

DeBono (1974) explains why he developed the game (p. 4):

> As a creative challenge I undertook to design the simplest possible classic game: a game that would be simpler than naughts and crosses [tic-tac-toe] and yet would be indeterminate and capable of being played with skill. The L-Game was the result. . . . The game requires a sense of spatial perception rather than sequential moves.

The game can be made easily from sheets of poster board of different colors. The cardboard should be heavy enough to ensure that board and pieces lie flat and do not curl.

*Rules of play.* The L-Game is for two players who alternate turns. A turn consists of moving one's own L-piece to any unoccupied position on the board. The L-piece

| Mathematics Problem | Strategy Game | Mathematical Puzzle |
|---|---|---|
| 1. Identify a real-world problem for which a solution is desired. | 1. Identify a strategy game for which a winning strategy is desired. | 1. Identify a mathematical puzzle for which a solution is desired. |
| 2. Look for and identify the mathematical relationships to be found in the problem. Translate these relationships into a suitable mathematical model. | 2. Look for and identify the mathematical relationships to be found in the game. Translate these relationships into a suitable mathematical model. | 2. Look for and identify the mathematical relationships to be found in the puzzle. Translate these relationships into a suitable mathematical model. |
| 3. Work out a solution for the model. | 3. Work out a solution for the model. | 3. Work out a solution for the model. |
| 4. Translate the solution of the mathematical model into the real-world terms of the problem. Apply the solution to the specific problem data. Evaluate the results. | 4. Translate the solution of the mathematical model into the terms of identifying winning situations and avenues of attack that lead to winning situations. Practice identifying winning positions, and practice applying strategies that lead to winning positions in actual game situations. Evaluate the results. | 4. Translate the solution of the mathematical model into the terms of the puzzle. Apply the solution to the specific puzzle data. Evaluate the results. |
| 5. Determine if the solution is sufficient and applicable. Does it satisfy the needs of the desired solution? If not, identify the part needing further investigation. Continue to develop the old mathematical model further or search for a new one. Return to step 3. | 5. Determine if the strategy developed is sufficient and applicable. Does it have both offensive and defensive components sufficient to ensure the degree of success desired? If not, identify the part of the strategy needing further investigation. Continue to develop the old mathematical model further or search for a new one. Return to step 3. | 5. Determine if the solution is sufficient and applicable. Is it simple, concise, and readily understood? If not, identify the part needing further investigation. Continue to develop the old mathematical model further or search for a new one. Return to step 3. |

Fig. 1

may be rotated and/or flipped before being placed back on the board. Its new board position *must* differ from its previous position by at least one square. After placing the L-piece, the player may, if so desired, move one (but not both) of the neutral pieces to a vacant square. The game ends when one player cannot make a legal move; the player last to move is the winner. Draws are possible, either by agreement or when each player repeats the same moves three times in succession.

Introducing a strategy game into the high school mathematics classroom for the purpose of problem-solving instruction requires some care. Virtually all students have played strategy games, but only a few have attempted any serious study of one. Initially, the problem-solving situation must be spelled out for them, namely, that it is their task to develop a winning strategy for the game. It may take the class several weeks to proceed through the task. They are certain to rely on the teacher for guidance, especially for leading them to the mathematics of the game.

With the L-Game, students should discover quickly that they need to identify the winning positions (those that leave the opponent with no legal move—see fig. 3) in order to develop a winning strategy. The game is small enough for a full analysis of this aspect to be well within the students' grasp.

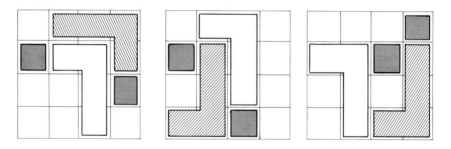

Fig. 3. Some winning positions: Stripe has no legal move.

In the analysis process the student will encounter and use numerous mathematical concepts—for instance, counting techniques, to identify the 18 368 possible positions, and equivalence classes, permitted by board symmetry, to identify the 2296 classes of 8 equivalent positions each. Of these classes, 1261 are draws (neither player can force a win from these positions), 1006 are wins for the player who is first to move from the position, and 29 are wins for the player who is second to move from the position (15 of these are immediate wins for the second player, since the first player is unable to move). Armed with these 29 positions, you need only look one move ahead to select a move that will prevent your opponent from moving into one of the 29 positions. This, of course, is a defensive strategy; the problem-solving search for an offensive strategy must continue. (See Truran [1976] for an excellent offense analysis of the game that uses the concept of translation. It is well within the grasp of a high school student.)

This should demonstrate that the search for a winning game strategy is, indeed, a genuine problem-solving experience in mathematics. Word problems typically found in high school algebra textbooks do not reflect this genuine problem-solving experience. Krulik (1977) prefers to label them *exercises*—''exercises in translating given relationships into the symbols and operations of algebra. . . . It would be erroneous to assume that through them students have been exposed to problem solving'' (p. 650). Through strategy games, however, students *are* exposed to problem solving. (See Krulik [1977] for further discussion of the use of strategy games for problem-solving instruction.)

## A Mathematical Puzzle

A square cake of uniform height is iced evenly on its top and on its four sides. Describe how to cut the cake into five pieces (vertical cuts only) such that each piece has the same amount of cake *and* the same amount of icing.

A mathematical puzzle chosen for problem-solving instruction in the classroom and presented to the class as a whole must have two essential properties: practically all students must feel that the problem is within their grasp, and the best students must find that its solution requires some effort. This gives the whole class an opportunity to experience a sustained problem-solving effort, allowing for a variety of approaches and, it is to be hoped, a variety of solutions.

The cake puzzle generally sends students in two different directions to search for a solution. The majority attempt an intuitive geometric solution; a handful (usually only the better students) try to set up a system of equations to solve. A nice feature of this problem is that the naive approach works best; sophisticated algebraic tools get too messy. Very often the first student to solve the problem is one whose success with verbal exercises has been less than satisfactory.

An intuitive geometric approach may proceed as follows: a square is sketched by the problem solver, who mentally notes the three-dimensional nature of the cake. He or she makes several symmetrical divisions of the square in the hope of getting lucky and stumbling on a solution. While exploring these guesses, the student observes that the cake's height is insignificant. Dividing the cake into five pieces of equal volume reduces to dividing the square into five pieces of equal area.

With this in mind, the would-be problem solver comes up with a proposed solution such as that in figure 4—four pieces touching the sides and the fifth in the center. Although the five pieces have equal amounts of cake, the student quickly discovers that such a division does not equalize the icing. All five pieces must have an equal share of the sides if they are to have the same amount of icing. The proposed solution in figure 4 must therefore be abandoned.

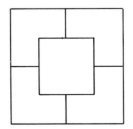

Fig. 4. An initial attempt at solution

The trick now is to divide the perimeter of the square into five parts of equal length. Divide each side into five sections of equal length and allocate four of these sections to each of the five pieces of cake (fig. 5).

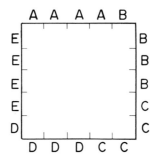

Fig. 5. Allocating the sides of the square

All that remains is to divide the interior of the square into five pieces of equal area. Superimposing a 5 × 5 square grid simplifies the process (fig. 6). After a little experimenting, a solution is found.

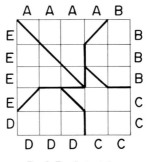

Fig. 6. The first solution

An algebraic approach is much less intuitive but, nevertheless, it does get the job done. Care must be taken to choose a starting point from which a solution is possible—by no means an easy task! Several false starts are typical. One starting position that works is given in figure 7. (I am indebted to Gary Prok for showing me this solution. He worked out the details while a freshman at the Ohio State University.)

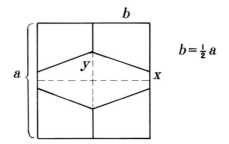

Fig. 7. Starting point for approach using algebraic tools

Symmetry permits one to focus on a quarter of the square (fig. 8).

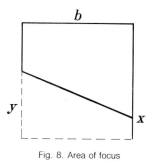

Fig. 8. Area of focus

The first equation is obtained from equating volumes. Let $h$ represent the height of the cake. The equation obtained is

$$4\left(\frac{y + x}{2}\right)bh = \left[\frac{(b - y) + (b - x)}{2}\right]bh.$$

Solving for $x$ produces

$$x = \frac{2}{5}b - y. \tag{1}$$

In the second equation the surface areas are equated as follows:

$$4\left[\left(\frac{y + x}{2}\right)b + xh\right] = \left[\frac{(b - y) + (b - x)}{2}\right]b + bh + (b - x)h.$$

Isolating $x$, we see that

$$\left(\frac{5}{2}b + 5h\right)x = -\frac{5}{2}yb + b^2 + 2bh.$$

Using (1) to substitute for $x$ and then simplifying, we get

$$5hy = 0.$$

So $y = 0$, and $x = \frac{2}{5}b$, yielding the picture in figure 9.

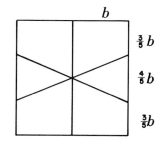

$\frac{3}{5}b$

$\frac{4}{5}b$

$\frac{3}{5}b$

Fig. 9. Initial solution produced by equations

Here, of course, the fifth piece is really two pieces "joined" at a single point, not quite what the problem calls for. A minor adjustment, trading congruent triangles, yields a satisfactory solution (fig. 10).

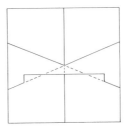

Fig. 10. Final solution

It is clear that the processes used to solve a mathematical puzzle can be quite different from those used to solve a routine exercise. With proper guidance, students will encounter genuine mathematical experiences while attempting the solution of a mathematical puzzle.

## The Question of Transfer

When recreational mathematics is used for problem-solving instruction, the transfer of skills to genuine problems is a question of concern. We do know that students enjoy recreational mathematics; this is certainly also true of strategy games. Students do study and learn problem-solving skills that they need to be effective in playing strategy games. They willingly take on a game as a challenge, and long after class is over, they work at it hour after hour. Often they will also spend hours working on a mathematical puzzle—although perhaps not with the same sustained effort that they typically put into strategy games.

Consider for a moment the following statements:

1. Skills acquired under enjoyable learning conditions are *usually* retained for long periods of time.

2. Skills acquired under unpleasant learning conditions are *usually* forgotten after a short-term goal has been reached. (Typical short-term goals are the completion of homework assignments, tests, and final examinations.)

3. As a precondition to transfer, skills acquired in one problem-solving situation must be retained at the time one confronts a different problem-solving situation requiring those skills in its solution.

Krulik (1977) observed that "there is little evidence to indicate that the ability to solve these exercises [i.e., word problems] offers any transfer to real problem-solving situations" (p. 651). One reason for this phenomenon is that students generally consider that learning to solve such exercises is an unpleasant experience—one they would quickly like to forget. For transfer to take place, the initial and subsequent learning experiences should be at the very least *not* unpleasant, but preferably enjoyable. The use of recreational mathematics provides such a learning environment.

However, transfer is not automatic simply because an environment conducive to transfer is present. The opportunity to transfer must also be present. Certainly, the similarity of steps found in figure 1 indicates that transfer will take place when the opportunity is there. Unfortunately, sufficient opportunities rarely occur by chance, and so the teacher must create them. Real-world problems must be carefully integrated with problems from recreational mathematics. The guidance of the teacher is initially strong, but as skills are strengthened it gradually weakens. Under these conditions, classroom practice confirms that transfer does take place. Thus, recreational mathematics provides a very effective means of involving students in problem solving.

## Suggested Games for Classroom Use

Here are a few suggestions for strategy games to try in the classroom.

**Tic-tac-toe.** A wealth of mathematics here, especially when one also examines several of the game variations. See Schuh (1968, pp. 60–100; 144–50) for a complete analysis of the game.

**Nim.** A known winning strategy uses the binary system, one unlikely to be discovered by a high school student. Nevertheless, other strategies are possible though perhaps not as complete. The binary strategy and proof are easy to follow. See Schuh (1968, pp. 60–100; 144–50) or Gardner (1959, pp. 73–83; 151–61) for the analysis.

**Hex.** Use board sizes of 2 × 2 cells up to 5 × 5 cells. The standard 11 × 11 board will probably never be completely solved. See Gardner (1959, pp. 73–83; 151–61) for further discussion.

**Plank.** When students grow tired of tic-tac-toe, Plank offers a delightful variation. See Sackson (1969, pp. 18–21) for the rules.

### REFERENCES

DeBono, Edward. "DeBono's L-Game." *Games & Puzzles* 30 (November 1974):4–6.

Gardner, Martin. *The "Scientific American" Book of Mathematical Puzzles and Diversions*. New York: Simon & Schuster, 1959.

Hunsucker, John L. "Recreational Mathematics for Teachers." *American Mathematical Monthly* 84 (January 1977):56.

Krulik, Stephen. "Problems, Problem Solving, and Strategy Games." *Mathematics Teacher* 70 (November 1977):649–52.

Krulik, Stephen, and Jesse Rudnick. *Problem Solving: A Handbook for Teachers*. Boston: Allyn & Bacon, 1980.

Sackson, Sid. *A Gamut of Games*. New York: Castle Books, 1969.

Schuh, Fred. *The Master Book of Mathematical Recreations*. New York: Dover Publications, 1968.

Schaaf, William L. *A Bibliography of Recreational Mathematics*. 4 vols. Reston, Va.: National Council of Teachers of Mathematics, 1970, 1970, 1973, 1978.

Trigg, Charles W. "What Is Recreational Mathematics?" *Mathematics Magazine* 51 (January 1978):18–21.

Truran, Trevor. "De Bono's L-Game." *Games & Puzzles* 47 (April 1976):17.

# 18

# *The Theme of Symmetry in Problem Solving*

## Gerald A. Goldin
## C. Edwin McClintock

**M**ANY mathematical problems can be solved using the *symmetry* that is intrinsic to the problem structure. Frequently the detection of hidden symmetry is the crucial step in insightful problem solving that enables the student to disregard irrelevant information and focus on the essential elements of the problem. In this essay we shall examine problems having different kinds of symmetry and discuss ways of recognizing symmetry when it is present in the structure of a problem. Our purpose is to encourage the use of symmetry in the solution process.

## Symmetry and the Study of Mathematics

One reason for stressing symmetry in teaching problem solving is its importance to many different branches of mathematics, as illustrated in the examples that follow.

In geometry a plane figure is said to have *line symmetry* (or *reflection symmetry*) if there is a line through which the figure can be reflected so that it is congruent to itself (fig. 1). A plane figure has *point symmetry* (or *rotation symmetry*) if there is a point about which the figure can be rotated so that it is congruent to itself (fig. 2). When point symmetry exists for rotation by $360°/n$, where $n$ is a whole number greater than 1, the plane figure has $n$-*fold* rotation symmetry. The Motion Geometry series by the University of Illinois Committee on School Mathematics (1969) contains a discussion of the concept of symmetry in geometry that is oriented toward junior high school students or better students in grades 5 and 6.

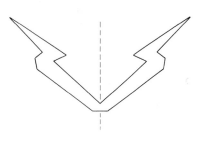

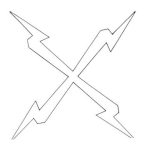

Fig. 1. Line symmetry                    Fig. 2. Point symmetry

By analogy with plane figures a three-dimensional figure has reflection symmetry when it possesses a *mirror plane* and rotation symmetry when it has an *n*-fold *axis of rotation*. It is easy to see, for example, that a cube has three 4-fold axes that are mutually perpendicular. It is perhaps less obvious that it also has four 3-fold axes (see fig. 3) and nine different mirror planes.

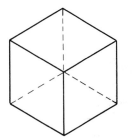

Fig. 3. A 3-fold axis of symmetry for a cube. The axis is perpendicular to the plane of the paper.

Symmetry in geometry can often be helpful in motivating the use of certain techniques in proofs. For instance, in the classical proof that the base angles of an isosceles triangle are congruent, a line segment is constructed from the apex *A* to the midpoint *D* of the base *BC*, as in figure 4. $\triangle ABD \cong \triangle ACD$ because the three pairs of sides are congruent; thus $\angle A \cong \angle B$. It may be suggested to the student that the construction of line segment *AD* is natural because it is a line of symmetry for the figure. A modern proof of the same theorem asserts that $\triangle ABC \cong \triangle ACB$, that is, that the original triangle is congruent to the *reflected* triangle; this proof uses the symmetry still more directly.

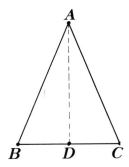

Fig. 4. Classical proof that the base angles of an isosceles triangle are congruent

Symmetry can also provide valuable mathematical insights into arithmetic, elementary algebra, and advanced high school algebra. For example, multiplication by −1 can be regarded as a reflection of the number line as shown in figure 5. This concept lends a natural interpretation to the equation $(-1)(-1) = +1$ as a representation of the fact that two successive reflections restore the number line to its original position. We thus obtain a visual representation of the properties of a product of several positive and negative numbers—the product is the number whose sign (direction) is in accordance with the number of reflections and whose magnitude equals the product of the magnitudes of the factors.

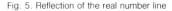

Fig. 5. Reflection of the real number line

The complex number $a + bi$ (where $i^2 = -1$) is located by a pair of Cartesian coordinates $(a, b)$ in the complex plane. Multiplying by −1 gives $(-1) \cdot (a + bi) = -a - bi$, which is represented by the coordinates $(-a, -b)$. This point is the same distance from the origin as $(a, b)$ but in the opposite direction. The operation of multiplying by −1 therefore *rotates* a point by 180° about the origin (fig. 6). The *complex conjugate* of the complex number $a + bi$ is defined as $a - bi$; this is symbolized by writing $\overline{a + bi} = a - bi$. When we compare the Cartesian coordinates $(a, b)$ and $(a, -b)$, we observe that the distances from the vertical axis are equal in magnitude and direction, whereas the distances from the horizontal axis are equal in magnitude but opposite in direction. Thus the operation of complex conjugation *reflects* a point through the horizontal (real) axis (fig. 7).

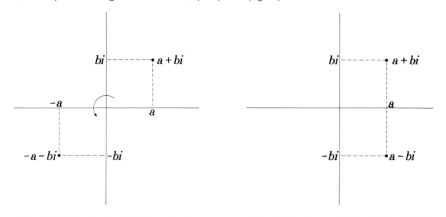

Fig. 6. Multiplication by −1 rotates points in the complex plane.

Fig. 7. Complex conjuation reflects points in the complex plane.

The symmetries of geometric figures provide a natural introduction to group theory and abstract algebra. Symmetry is also important to many applications of mathematics. Modern physics has been deeply influenced by the discovery that the laws of nature are *not* invariant under spatial reflection. In chemistry the possible symmetries of crystals are described by thirty-two different mathematical groups (crystallographic "point groups"), and the symmetry of molecular configurations simplifies the problem of predicting electron densities (fig. 8). Art (see fig. 9) and architecture have likewise been influenced by the mathematics of symmetry. Gamow (1947) includes a discussion of symmetry in the physical universe, and Gardner (1964) explores symmetry in physics, chemistry, music, and art. Gardner is also noted for his fascinating discussions of mathematical games and puzzles, many of which embody symmetry; see, for example, Gardner (1959, 1966), recommended for strong high school students.

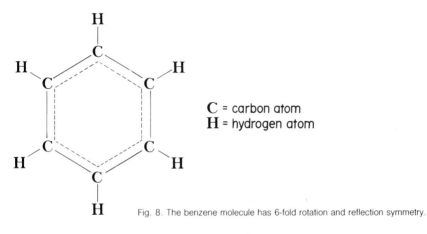

C = carbon atom
H = hydrogen atom

Fig. 8. The benzene molecule has 6-fold rotation and reflection symmetry.

Fig. 9. M. C. Escher's *Circle Limit IV* exhibits 3-fold rotation and reflection symmetry (Escher Foundation—Haags Gemeentemuseum—The Hague).

Thus an understanding of symmetry in mathematics is likely to deepen one's insight into almost every branch of the subject and into the relationships among different mathematical subject areas. In our view, the exploitation of symmetry in problem solving can systematically develop such understanding and lay the foundation for a sound mathematical intuition that will prove valuable in more advanced study.

## Symmetry in Insightful Problem Solving

Noninsightful problem solving can be a difficult, tedious process. At best it may lead to algebraic or numerical answers that carry no particular significance, thereby encouraging the rote memorization of algorithms or formulas. Most commonly it generates failure, frustration, and a lasting dislike for mathematics. By contrast, we associate insightful problem solving with elegant, enjoyable solutions, pleasing shortcuts, and little or no need for rote learning.

Symmetry can be the key to insight into the structure of many mathematical problems. We shall offer three examples. The first is a well-known problem requiring just a single observation. In the latter two problems, attention to symmetry yields fairly deep insights.

**Problem 1.** A rectangle is inscribed in a quadrant of a circle as shown in figure 10. Given that the length of segment $OA$ is 5 and that of segment $AP$ is 1, find the length of the diagonal $AC$.

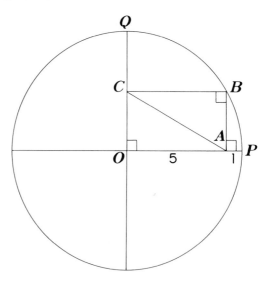

Fig. 10. The inscribed rectangle problem

In attempting to solve this problem, the student usually notices that finding $\overline{OC}$ would permit the use of the Pythagorean theorem to find $\overline{AC}$. It may be observed that $\overline{OC} = \overline{AB}$. The question then becomes how to use the given data $\overline{OA} = 5$ and $\overline{AP} = 1$ to find $\overline{AB}$. Here many students become lost drawing auxiliary lines such as

$BP$ and $BQ$, labeling angles, and trying to use trigonometry. The insight, however, is that the reflection symmetry of the rectangle permits the required diagonal $AC$ to be placed in congruence with its equivalent diagonal $OB$. The latter, of course, is a radius of the circle—immediately observed to have length 6.

The following problem offers the possibility of several solution strategies, including an elegant solution based on symmetry.

**Problem 2.** In figure 11, a TV game show contestant must run from her initial starting position $A$ to a long table $CD$ that is covered with chocolate cream pies. The table is 13 meters long and is placed 5 meters from $A$. After picking up a pie from the table, the contestant is to race to her partner, who is 8 meters from the table at $B$, and give him a faceful of pie. What is the shortest distance in which she can accomplish this feat?

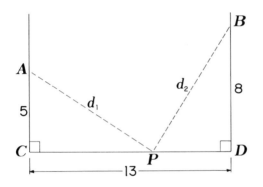

Fig. 11. The chocolate cream pie problem

A brute-force strategy for solving this problem is available from calculus. Letting $x = \overline{CP}$ and $y = d_1 + d_2$ in figure 11, we have from the Pythagorean theorem $y = (5^2 + x^2)^{1/2} + (8^2 + [13 - x]^2)^{1/2}$. Now it is a matter of computing the first derivative $dy/dx$, equating it to zero, and solving for $x$. This is an excellent example of non-insightful problem solving through following an algorithmic procedure.

A more interesting strategy that does not depend on calculus is to attempt a systematic trial-and-error, or successive approximation, approach. The following chart contains some data derived from the conditions of the problem with a hand calculator.

| $\overline{CP}$ | $\overline{PD}$ | $d_1$ | $d_2$ | $d_1 + d_2$ |
|---|---|---|---|---|
| 0 | 13 | 5.000 00 | 15.264 34 | 20.264 34 |
| 2 | 11 | 5.385 16 | 13.601 47 | 18.986 63 |
| 4 | 9 | 6.403 12 | 12.041 59 | 18.444 71 |
| 6 | 7 | 7.810 25 | 10.630 15 | 18.440 40* |
| 8 | 5 | 9.433 98 | 9.433 98 | 18.867 96 |
| 10 | 3 | 10.440 31 | 8.544 00 | 18.984 31 |
| 13 | 0 | 13.928 39 | 8.000 00 | 21.928 39 |

\* approximate location of minimum

The near equality of the values of $d_1 + d_2$ for distances $\overline{CP}$ of 4 and 6 suggests that the next approximation might appropriately be $\overline{CP} = 5$. This indeed yields a smaller value for $d_1 + d_2$, namely, 18.384 78. But it has not been proved that this is the minimum distance, and the symmetry intrinsic to the problem has not yet been exploited.

The question of finding the shortest distance $\overline{APB}$ may eventually suggest transforming this path into a straight line—a transformation that can conceivably be accomplished by reflecting the point $B$ through the line $CD$ to the *symmetrically conjugate* point $B'$. Since such a reflection preserves distances, the length of the path $APB$ must be the same as that of the path $APB'$, and this length is minimized by the straight line $AB'$ as indicated in figure 12. Now it is easily seen, using similar triangles, that $\overline{CP} = 5$. Thus the original problem has an absurdly simple and elegant solution once the concept of symmetry is used—if the pie table is lined with a mirror, the TV contestant runs toward her partner's mirror image until she can pick up a pie and does not use any mathematics at all.

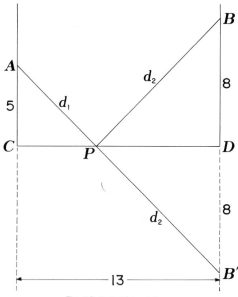

Fig. 12. Insightful solution

This construction may seem like just a trick, but in fact it is a profound result, having application in physics for determining the path traversed by a rebounding billiard ball or a reflected beam of light. It has as a consequence the law that the angle of incidence is equal to the angle of reflection.

Symmetry can assist with the insightful solution of problems from algebra as well as from geometry. The two preceding problems incorporated reflection symmetry; the next problem will introduce $n$-fold rotation symmetry.

**Problem 3.** Find the sum of the infinite series

$$\frac{1}{4} + \frac{1}{16} + \frac{1}{64} + \frac{1}{256} + \ldots$$

Of course it is well known that if $x$ is a real number with absolute valueless than one, the sum of the infinite geometric progression $x + x^2 + x^3 + x^4 + \ldots$ is equal to $x/(1 - x)$. This equation, or one like it, is often obtained in second-year algebra, and students are expected to remember the formula at least until the next examination. Thus our third problem can be solved by substituting $x = 1/4$ into the formula, from which we obtain the result that the series sums to $1/3$. However, this is scarcely insightful problem solving. Can the sum be *visualized,* without relying on algebraic symbol manipulation?

For $x = 1/2$ we have the related infinite series $(1/2) + (1/4) + (1/8) + (1/16) + \ldots$. It is possible to visualize this sum as a sequence of steps traversing an interval of unit length. By crossing $1/2$ of the interval, followed by $1/2$ of the remaining distance, and again $1/2$ of the remaining distance, ad infinitum, one approaches closer and closer to the endpoint 1 of the interval. Thus the sum of the series equals 1. Unfortunately, this method will not work for $x = 1/4$. Indeed, it is nearly impossible for most students (or most teachers!) to visualize the limit point of the sum $(1/4) + (1/16) + (1/64) + \ldots$ on an interval of unit length.

Another way to proceed is to call the desired sum $y$. We observe that all the terms of $y$ are included in the sum $(1/2) + (1/4) + (1/8) + (1/16) + \ldots$, which totaled 1. If we subtract out these terms the result is $1 - y = (1/2) + (1/8) + (1/32) + \ldots$. But this series is the result of *doubling* each term in the series for which the sum is desired; that is, $(1/2) + (1/8) + (1/32) + \ldots = 2 \cdot [(1/4) + (1/16) + (1/64) + \ldots]$. Thus $1 - y = 2y$, and $y = 1/3$.

Now we have obtained the answer without making use of the algebraic formula. Furthermore, we have the added satisfaction of having related the given series to a more familiar series. However, this method does not generalize to provide the sum for a series such as $(1/5) + (1/25) + (1/125) + \ldots$. Thus the search for still another approach is indicated.

If one is motivated by the desire to use symmetry, one can construct a concrete representation of the infinite series, as in figure 13. Imagine that three people are eating cake. The cake is cut into four congruent pieces, each person takes one piece $(1/4$ of the cake), and there is one piece left over. Out of courtesy, no one wishes to take the last piece; so this is again cut into four congruent pieces. Each person takes one piece $(1/4$ times $1/4$, or $1/16$ of the cake), leaving one piece remaining, which is again cut into four congruent pieces. This process is continued until only a microscopic crumb remains. Each person has thus eaten $(1/4) + (1/16) + (1/64) + \ldots$ of the cake. But there were *three* people, and by the symmetry of the process, each person ate an equal amount of cake—that amount must therefore have been $1/3$.

This method generalizes easily to the case of $n - 1$ people sharing the cake. The cake is initially cut into $n$ pieces, each person takes a piece, and one piece is left over. It is possible to visualize, by continuing the process, the result that

$$\frac{1}{n} + \left(\frac{1}{n}\right)^2 + \left(\frac{1}{n}\right)^3 + \ldots = \frac{1}{n - 1}.$$

With $x = 1/n$, this is equivalent to the original formula.

The same concrete representation makes it possible to visualize the finite sum $x + x^2 + x^3 + \ldots + x^m$, which equals $(x - x^{m+1})/(1 - x)$. For example, in figure 13, we can see that $(1/4) + (1/16) = [1 - (1/16)]/3$ and $(1/4) + (1/16) + (1/64) = [1 - (1/64)]/3$, thus leading to the more general equation

$$\frac{1}{n} + \left(\frac{1}{n}\right)^2 + \ldots + \left(\frac{1}{n}\right)^m = \left[1 - \left(\frac{1}{n}\right)^m\right]\bigg/(n - 1).$$

With $x = 1/n$, this is again equivalent to the original formula.

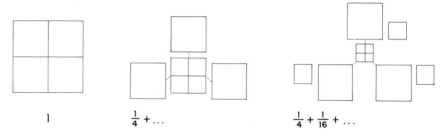

$1$        $\frac{1}{4} + \ldots$        $\frac{1}{4} + \frac{1}{16} + \ldots$

Fig. 13. Visualizing the sum of an infinite geometric progression

One might well ask how one arrives in the first place at the idea of cutting the cake into four pieces, or of letting three people share the cake. Clearly it is necessary, but not sufficient, for the problem solver to have a concept that there exists such a thing as symmetry and that it might be a good idea to use it when possible. In addition, it is important to look at the statement of the problem, particularly its givens, for clues that symmetry might be present. Here the first term in the given series, $1/4$, can be represented *asymmetrically* as one-fourth of the distance along a unit interval or *symmetrically* as any one of four equivalent portions of a unit quantity. Being predisposed toward the use of symmetry encourages the latter choice, and the continuation of this idea leads to the representation in figure 13. We shall return to the issue of the detection of problem symmetry later in this essay.

## Kinds of Symmetry in Mathematical Problems

A mathematical problem usually consists of a verbal statement summarizing the given information, the conditions or rules of procedure, and the goal or goals to be reached. Sometimes the verbal statement of the problem is accompanied by a diagram or physical representation; at other times this representation must be constructed by the problem solver. An important aspect of symmetry is the extent to which the symmetry is overt as opposed to hidden. We shall consider the following three possibilities: (a) symmetry is overtly present in the given representation of the problem; (b) the given representation requires modification or additional construction in order to display symmetry; and (c) the symmetry is hidden or not apparent in the given representation and a new representation must be constructed in order to display it. For any of these possibilities, the symmetry in question may be reflection symmetry, rotation symmetry, and so on, as described in the first section.

### Overt symmetry

Examples of overt symmetry have already been encountered in the proof that the base angles of an isosceles triangle are congruent (fig. 4) and in the problem of the inscribed rectangle (fig. 10). The triangle in figure 4 and the rectangle in figure 10 both possess overt reflection symmetry, which suggests the elegant solutions. However, the fact that symmetry is overt does not necessarily mean that a problem is easy, as the following example demonstrates.

**Problem 4.** Let $ABCD$ be a rectangle and let $M$ be the midpoint of side $AB$ (see fig. 14). Suppose that a perpendicular is erected to line segment $MC$ at $M$, intersecting side $AD$ at $P$. Prove that angle $BCM$ is congruent to angle $PCM$.

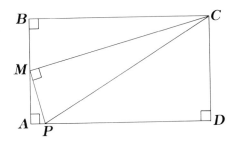

Fig. 14. The overt symmetry of the rectangle suggests auxiliary constructions.

The overt symmetry of the rectangle in figure 14 suggests the construction of the line of symmetry that passes through $M$, intersecting $CD$ at $N$ and $PC$ at $Q$ (fig. 15). Since $M$ is the midpoint of $AB$, $Q$ is the midpoint of $PC$. Since $\triangle PMC$ is a right triangle, $\angle QMC \cong \angle QCM$ (as is seen by the symmetry in fig. 16). But $\angle QMC$ is also congruent to $\angle BCM$, and the sketch of the proof is complete.

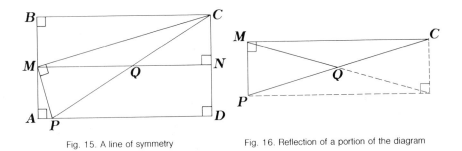

Fig. 15. A line of symmetry          Fig. 16. Reflection of a portion of the diagram

Overt symmetry can occur in problems other than geometry problems. Many mathematical games and puzzles have concrete representations possessing symmetry; these are ideal for instruction at the upper elementary or junior high school level. For example, the ticktacktoe grid displays 4-fold rotation symmetry as well as reflection symmetry, which permits various moves to be treated as equivalent (see fig. 17).

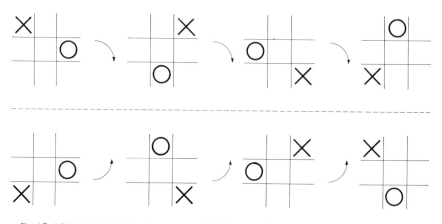

Fig. 17. 4-fold rotation and reflection symmetry in ticktacktoe: all the indicated situations are equivalent.

Problem 5 is a game described by Jacobs (1970) in which there is a winning strategy based on symmetry (fig. 18). Jacobs's book, "a textbook for those who think they don't like the subject," contains many additional illustrations of symmetry in mathematics.

**Problem 5.** A coin is placed at each vertex of a regular polygon. The two players take turns removing either a *single* coin or *two coins at adjacent vertices.* The winner is the player who removes the last coin. What is the winning strategy?

Here, too, the game board possesses overt rotation and reflection symmetry. The second player can always win by selecting the one or two coins exactly opposite the coin or coins removed by the first player, thus leaving the board in a pattern that has reflection symmetry (see fig. 19).

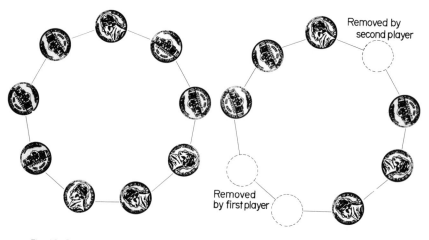

Fig. 18. Game board for problem 5                 Fig. 19. A winning strategy in problem 5

## Symmetry in a modified representation

For some problems the given representation requires modification or addition in order to display the symmetry, as in the problem of the chocolate cream pies (figs. 11 and 12). Suppose, for another example, that the game in problem 5 is presented to middle school students with a game board that has been deformed from the shape of a regular polygon, as in figure 20. Now the detection of the winning strategy may require recognition of the fact that the rules of the game are not affected if the allowed locations of the coins are rearranged in the shape of a regular figure. This is another way of saying that the network properties of figure 20 are equivalent to those of figure 18. The representation can be altered so that the symmetry becomes apparent.

Fig. 20. A modified version of problem 5

## Hidden symmetry

Finally we have the type of problem for which a new representation must be constructed in order to display the problem's symmetry. Problem 3 (the sum of the infinite series) provided an example of this situation, which is, of course, the most difficult of the three possibilities. Problem 6 is a well-known game that has hidden symmetry.

**Problem 6.** The numbers 1 through 9 are written on scraps of paper that are placed faceup between the two players. The players alternate in selecting a number. The winner is the first player to select exactly three numbers adding to fifteen.

This game is sometimes called *number scrabble.* Its symmetry is most easily understood by noting the correspondence of moves in number scrabble with moves in ticktacktoe. The magic square in figure 21 provides a new representation for number scrabble in which the symmetry is overt instead of hidden.

| 2 | 7 | 6 |
|---|---|---|
| 9 | 5 | 1 |
| 4 | 3 | 8 |

Fig. 21. Magic square displaying the isomorphism between number scrabble and ticktacktoe

Two problems of games (such as ticktacktoe and number scrabble) for which there is a one-to-one correspondence of legal steps or moves are said to be isomorphic. Isomorphic problems have the property that all symmetry that is present in one problem is present in the other (even though it may be overt in one problem and hidden in another); any strategy that can be applied to one problem can also be applied to the other. The next two famous problems are isomorphic to each other.

**Problem 7.** In the Tower of Hanoi problem, four concentric rings are placed in order of size, the smallest at the top, on the first of three pegs (see fig. 22). The object of the problem is to transfer all the rings from the first peg to the third in a minimum number of moves. Only one ring may be moved at a time, and no larger ring may be placed above a smaller one on any peg. (The problem generalizes to $n$ rings on three pegs.)

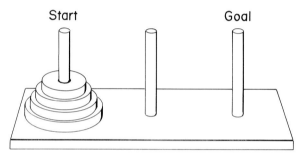

Fig. 22. The 4-ring Tower of Hanoi

**Problem 8.** In the Tea Ceremony a host, an elder, and a youth are performing four tasks, listed in ascending order of importance: stoking the fire, passing the cakes, serving the tea, and reciting poetry. Initially the tasks are all performed by the host, and the problem is to transfer all four tasks to the youth. Only one task—the least important a person is performing—may be transferred at a time, and no participant may accept a task unless it is less important than those he or she is already performing. How can the four tasks be transferred in the fewest possible steps?

The three people in the Tea Ceremony—host, elder, and youth—correspond to the three pegs in the Tower of Hanoi—the start peg, the middle peg, and the goal peg, respectively. The four tasks—stoking the fire, passing the cakes, serving the tea, and reciting poetry—correspond to the four rings, from the smallest to the largest. The tower of Hanoi problem and correspondingly the Tea Ceremony problem possess a symmetry that arises from *interchanging* the roles of the goal peg and the middle peg. Any sequence of moves that results in the four rings being transferred to the middle peg has a corresponding sequence in which the four rings are transferred to the goal peg. If the three pegs were set at the vertices of an isosceles triangle, with the start peg at the apex of the triangle, the symmetry of the problem would be apparent as a reflection about the altitude.

It is important to note the usefulness of isomorphic problems in detecting and describing the hidden symmetry of a problem and in facilitating the *transfer* of learning from one problem-solving situation to another.

*Forward-backward* symmetry is another type of symmetry that may occur in mathematical problems. For example, consider one variation of the famous "missionary-cannibal problem" given in problem 9.

**Problem 9.** Three missionaries and three cannibals are on one bank of a river. They have a rowboat that will hold at most two people. All the missionaries know how to row, but only one of the cannibals can row. How can the group cross to the other side of the river in such a manner that missionaries are never outnumbered by cannibals on either riverbank?

The complete set of legal moves for this problem is pictured in figure 23. Clearly, if one begins at the goal position instead of the initial position and works backward in-

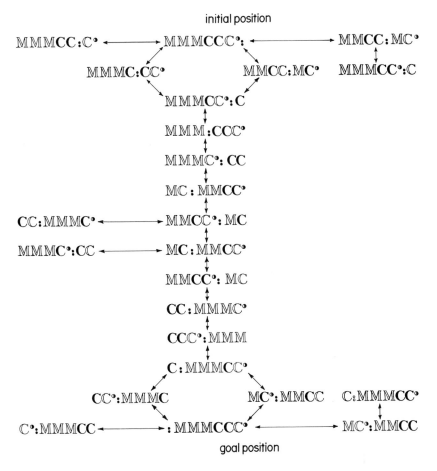

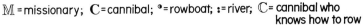

Fig. 23. Legal moves for the missionary-cannibal problem

stead of forward, the problem is identical in structure to the original. Of course it is not necessary to diagram all the legal moves to recognize the forward-backward symmetry—the symmetry follows immediately from the reflection symmetry in the description of the problem. Crossing from the right bank of the river to the left is not essentially different from crossing left to right.

When forward-backward symmetry is present, the solution of the problem is no easier in one direction than in the other. However, the symmetry can still be exploited by the problem solver. Any strategy that will generate a series of moves working forward will generate a corresponding series of moves working backward. An often useful technique is to develop several steps forward from the initial configuration and then to work backward several equivalent steps from the goal configuration, looking for a connection in the middle. The missionary-cannibal problem yields readily to this method. Forward-backward symmetry is also present in the Tower of Hanoi and Tea Ceremony problems.

To sum up, we have observed that problems, like geometric figures, may be endowed with reflection or rotation symmetry. The symmetry of a problem may be overtly present in an accompanying representation, the representation may need modification in order to display the symmetry, or the symmetry may be hidden and an entirely new representation may be necessary in order to detect it. Besides the usual reflection and rotation symmetry, problems may have forward-backward symmetry in which the goal situation is treated as given and the given situation as the goal. Isomorphic problems share reflection, rotation, and forward-backward properties of symmetry, but the symmetry that is hidden in one version may be overt in another. All these properties of symmetry can be useful in insightful problem solving.

## Teaching the Detection and Exploitation of Symmetry in Problem Solving

The use of symmetry in problem solving usually does not conform to any routine algorithm. Polya (1965) calls attention to problem symmetry through the "principle of nonsufficient reason," which he expresses thus: "No one should be favored of eligible possibilities among which there is no sufficient reason to choose" (p. 161), or "We expect that any symmetry found in the data and condition of the problem will be mirrored by the solution" (p. 161). Polya's books (1945, 1965) also contain many examples of problems that incorporate symmetry. To teach these ideas effectively, it is important to shift away from an exclusive orientation toward correct answers and to emphasize the *processes* of problem solving. As we have seen, problems with symmetry often lend themselves to noninsightful as well as insightful problem solving; to foster the latter, we must pay attention to the solution process.

An important step is the creation of a framework of teaching objectives that stress the use of particular processes. The list of objectives in figure 24 is proposed as an example suitable for a spiral approach; that is, each category of objectives may be taught at one level of sophistication and reentered at a higher level. It is our view that these objectives should be made explicitly known to the students, and classroom activities should be organized to achieve them.

This essay illustrates most of the concepts listed in figure 24. It is generally helpful

---

**Teaching Objectives for the
Theme of Symmetry in Problem Solving**

1. Descriptive features of rotation and reflection symmetry

    Detection of symmetry in geometric figures (*n*-fold rotation and reflection symmetry, plane and solid figures); classification of geometric figures by type of symmetry; construction of figures to illustrate different types of symmetry; detection of patterns of symmetry in arithmetic and algebra (e.g., symmetry in addition and multiplication tables, Pascal's triangle, the real number line, and the complex plane)

2. Groups of symmetry transformations

    Symmetry transformations and congruence; the concept of successive transformations; group properties (closure, associative property, identity element, inverse elements)

3. Detection of overt symmetry in problem representations

    Identification of reflection, rotation, and forward-backward symmetry when it is overtly present in the representation of a problem (e.g., ticktacktoe, the Tower of Hanoi, the missionary-cannibal problem, geometric problems of proof with symmetric figures)

4. Use of overt symmetry in problem solving

    Game strategies using symmetry and game situations equivalent by virtue of symmetry; construction of lines of symmetry in geometry; use of the ''working backward'' technique

5. Problem isomorphisms

    The concept of isomorphic problems; detecting isomorphisms in problems; inventing isomorphic problems; applying strategies learned in one problem to an isomorphic problem; symmetry in isomorphic problems

6. Detection and use of hidden symmetry in problem solving

    Inferring symmetry from a problem statement; construction of problem representations that display symmetry; modification of representations to make symmetry more apparent; the use of isomorphic problems to make hidden symmetry overt

---

Fig. 24

to stress the discovery or detection of symmetry patterns by the students rather than to point out these patterns too early. It is also worthwhile to relate newly detected symmetry patterns to patterns discovered in previously studied problems.

## Concluding Remark

In his study of the structure of mathematical abilities in children, Krutetskii (1976) identifies three important components: (1) the ability for rapid and broad generalization; (2) the ability to curtail the process of mathematical reasoning; and (3) the reversibility of the mental process. Krutetskii's interesting analysis includes an extensive system of experimental problems that will be of interest to many teachers.

The theme of symmetry in problem solving, although not stressed by Krutetskii, provides a means of fostering the development of these components of ability. Isomorphic problems permit and encourage the use of generalization. The shortcuts provided by treating as equivalent those problem situations that are related by symmetry present opportunities for the curtailment of reasoning. Problems with forward-backward symmetry can be used to illustrate the change from a direct to a reverse train of thought. In short, the teaching of problem solving using symmetry may help the student develop a more powerful mathematical intuition that will be helpful in later years.

## REFERENCES

Gamow, George. *One, Two, Three, . . . , Infinity.* New York: Viking Press, 1947.

Gardner, Martin. *The Ambidextrous Universe.* New York: Basic Books, 1964.

————. *New Mathematical Diversions from Scientific American.* New York: Simon & Schuster, 1966.

————. *The Scientific American Book of Mathematical Puzzles and Diversions.* New York: Simon & Schuster, 1959.

Jacobs, Harold R. *Mathematics: A Human Endeavor.* San Francisco: W. H. Freeman & Co., 1970.

Krutetskii, V. A. *The Psychology of Mathematical Abilities in Schoolchildren.* Edited by Jeremy Kilpatrick and Izaak Wirszup and translated by Joan Teller. Chicago: University of Chicago Press, 1976.

Polya, George. *How to Solve It.* Princeton, N.J.: Princeton University Press, 1945.

————. *Mathematical Discovery: On Understanding, Learning, and Teaching Problem Solving,* vol. 2. New York: John Wiley & Sons, 1965.

University of Illinois Committee on School Mathematics. Motion Geometry series. New York: Harper & Row, Publishers, 1969.

*19*

# Some Thoughts on Teaching for Problem Solving

## Mary Grace Kantowski

TEACHING for problem solving differs from all other aspects of mathematics instruction. Most teachers would agree that planning instruction to help students improve their ability to solve difficult, nonroutine problems is the most challenging task facing them in the mathematics classroom.

Instruction for problem solving must be approached as a system. When the study of any system is begun, it is important to establish definitions and assumptions. The system for problem-solving instruction discussed in this essay depends on a definition of *problem* and on three assumptions.

### The definition of problem

A problem is a situation for which the individual who confronts it has no algorithm that will guarantee a solution. That person's relevant knowledge must be put together in a new way to solve the problem. The following examples should help us make a distinction between real *problems* and those generally found at the end of textbook sections, better known as *exercises*.

1. Steers sell for $25 a head and cows for $26 a head. A farmer has $1000 to spend and must spend it all on cattle. How many cows and how many steers should he buy?

2. Steers sell for $25 a head and cows for $26 a head. A farmer bought 14 steers and 25 cows. How much did he spend in all?

3. The sum of three consecutive integers is 279. Find the integers.

4. Find two sets of three consecutive primes such that the sum of the numbers of one set is the reverse of the sum of the numbers of the other. (Prielipp 1972)

Clearly, algorithms are available to find solutions for 2 and 3; these are exercises. For most of us, 1 and 4 are, on first contact, nonroutine problems.

### Some assumptions

Our system for problem-solving instruction is based on three assumptions:

1. Problem solving, in some form, is for everyone. When and if textbooks do include problems, they are often labeled "For the Experts." *All* mathematics students, regardless of ability level, deserve to share the pleasures of problem solving!

2. For most students, expertise in problem solving doesn't just happen. It is the result of a combination of carefully planned instruction and experience in solving a variety of problems.

3. Finally, problem solving cannot be learned in a crash course. For most students, the ability to solve problems develops *slowly* over a long period of time.

Now we are ready to look at what to teach when teaching for problem solving, as well as how to approach instruction.

### What to teach

There are at least two essential components of successful problem solving: (1) knowing something of the relevant mathematics and (2) knowing what to do with what is known. It is important that students have mathematical skills beyond computational skills. For example, recognizing perfect squares and perfect cubes or Pythagorean triples is often crucial to the solution of a problem. Polya (1973, p. 5) suggests that experience in solving a variety of problems is one of the most important factors in the development of problem-solving ability. One important outcome that could result from solving a variety of problems in geometry is an intuitive feeling for certain geometric relationships.

To illustrate, let us look at the following problem:

> Given two equilateral triangles that measure 12 cm on a side. The triangles are positioned to form a regular six-pointed star. What is the area of the overlapping figure?

The overlapping figure is a hexagon (see fig. 1) that can, of course, be easily divided into six equilateral triangles.

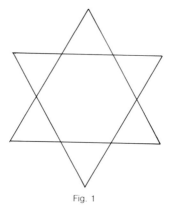

Fig. 1

If a student has developed a good feeling for the relationship that exists between an equilateral triangle and a hexagon, the diagram in figure 2 could emerge, making the solution intuitively clear. Since the star is regular, it is clear from the figure that the side of each of the small triangles is one-third that of the original triangle.

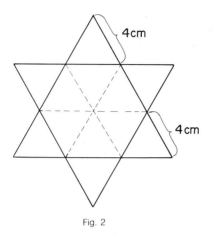

Fig. 2

Thus, the area of the hexagon, $K_h$, is

$$K_h = 6\left(\frac{4^2\sqrt{3}}{4}\right) \text{cm}^2$$

$$= 24\sqrt{3} \text{ cm}^2.$$

The proof is easy once the solution is *seen*. But the student must have a sense of the relationships that exist and, of course, know the formula for the area of an equilateral triangle given its side.

Similarly, solving problems founded in number theory gives students an intuition for the properties of numbers. Such intuition often enables them to make conjectures about a possible solution to a given problem if, indeed, such a solution exists. An example, taken from the *Stanford Mathematics Problem Book* (Polya and Kilpatrick 1974, p. 19), illustrates how a feeling for the properties of numbers can truly simplify a problem that looks quite difficult at first glance:

> A certain make of ball point pen was priced at 50 cents in the store opposite the high school but found few buyers. When, however, the store had reduced the price, the whole remaining stock was sold for $31.93. What was the reduced price?

A problem solver who has developed good number sense will immediately "see" that 31 is a factor of 3193.

An experienced problem solver would also recognize that 103 is *prime*—giving a very quick solution to the problem! Although the solution seems quite simple once it has been found, the skills of recognizing 31 as a factor of 3193 and of noting that 31 and 103 are prime are crucial to finding the solution.

In addition to a knowledge of the necessary mathematics, some ideas about what to do with this knowledge are necessary in teaching for problem solving. Every teacher has encountered students who know the facts but seem helpless when faced with problems that require more than a simple application of a formula. We have all heard students say, ''I see what you did, but how did you think of doing that?'' or ''I can always follow your solutions, but I would never think of doing some of those things myself.'' Something else—a skill different from computational skill—is a necessary component of successful problem solving. The ability to think of what to do is as central as knowing the facts or having the necessary computational facility. Some rules of thumb or ideas on what to try in a given situation have been suggested by Polya (1973, pp. xvi–xvii). He notes four phases in the solution of a problem and proposes a set of heuristic questions related to each of the phases. Recent research studies (Kantowski 1977; Schoenfeld 1979) have shown that the use of heuristics such as those suggested by Polya's questions are clearly related to success in problem solving.

### Some thoughts on how to . . .

What a student should know is one important element to be considered in planning instruction in problem solving. Just as significant, however, are two other essential components, the role of the teacher and the organization of instruction. In an effective system these two components are integrated in practice. The teacher's role should change from *model* to *prosthesis* to *problem provider* to *facilitator* as the ability of the students to solve problems develops. Students enter mathematics classes at different levels of proficiency in problem solving. In reality, these levels are not discrete but form a continuum ranging from a complete lack of understanding of what problem solving is all about to some competence in solving the most complex nonroutine problems. To facilitate the characterization of the teacher's role, let us consider students at four points on the continuum corresponding to four different levels of problem-solving ability. The teacher's function is different at each level with less involvement in the actual problem solving as the students' capabilities increase. Figure 3 indicates the changing characteristics of students in their problem-solving development and the suggested role of the teacher at each level.

---

Characteristics of students at each level of problem-solving development and the role of the teacher at each level

First level—

Students have little or no understanding of what problem solving is, of the meaning of strategy, or of the mathematical structure of the problem. Most students at this level do not know where to begin to solve a nonroutine problem.

Teacher assumes the role of model.

Second level—

Students understand the meaning of problem solving, of strategy, and the mathematical structure of a problem. They are able to follow someone else's solution and can often suggest strategies to be tried for problems similar to those they have seen before. Although they will participate actively in group problem-solving activities or instructional episodes, many feel insecure about independent problem solving.

*Teacher* acts as *prosthesis,* or crutch.

Third level—

*Students* begin to feel comfortable with problems. They suggest strategies different from those they have seen used. They understand and appreciate that problems may have multiple solutions and that "no solution" may be a perfectly good solution.

*Teacher* becomes *problem provider.*

Fourth level—

*Students* are able to select appropriate strategies for most problems encountered and are successful in finding solutions much of the time. They show an interest in elegance and efficiency of a solution and in finding alternate solutions to the same problem. They suggest variations of old problems and are constantly searching for novel problems to challenge themselves and others.

*Teacher* serves as *facilitator.*

Fig. 3

*First level* . In approaching the instruction of novice problem solvers—students at the first level—the teacher should assume the role of *model* . A carefully selected problem could be solved from scratch, perhaps using Polya's heuristic questions. Dead ends, unfruitful excursions, and inefficient first solutions should be included in modeling a solution so that students will become aware that often problems are not solved neatly right away. Solutions often require many false starts as well as numerous revisions before an elegant or efficient one is obtained. (Many students are aware of this, but it is gratifying for them to see that even experienced problem solvers do not always go through the maze of the solution process without running into a host of obstacles!) Most important, the instructor should model the "cleaning up" process, which corresponds to one aspect of Polya's *looking back* phase, where the solution is studied, extraneous steps are removed, and alternative solutions to the same problem or new problems suggested by the solution are put forth.

Let us look at a sample modeling of a solution for the well-known triangular number problem that could be presented to students at the first level:

A number that may be represented by the pattern below is called a triangular number. The first four are shown. What is the fiftieth triangular number?

Since most students at the first level would have no idea where to start or even what is being asked, the teacher might point out that the first triangular number is represented by one row of dots with one dot in a row, the second triangular number is represented by two rows with two dots in the second row, and the third triangular number by three rows with three dots in the third row, and so forth. The students should recognize the pattern and may even conjecture that the fiftieth triangular number could be found by counting the dots of the expanded triangle with fifty dots on a side. In fact, that gives a perfectly correct solution (although not a very efficient one, to be sure!), and if the group does not propose it, the teacher should note it as a possibility to make the point that if all else fails, a persistent problem solver will "grind it out." If no one suggests that there must be an easier way, the teacher might then model "looking back" to try to find a cleaner solution.

First the dots could be represented in symbol form:

$$T_1 = 1$$
$$T_2 = 1 + 2 = 3$$
$$T_3 = 1 + 2 + 3 = 6$$

Such a representation indicates that the fiftieth triangular number could be found by finding the sum of the first fifty positive integers. A logical next step would be to ask for *any* triangular number, perhaps the 187th or the 5000th or the 1 000 000th. Adding the first 5000 integers (even with the help of a calculator) would be unrealistic. It is plain that what is needed is to find the *k*th triangular number by using *k* somehow. As primitive as this notion is to a mathematician, the idea of finding a *k*th term, given *k*, is not an easy one for beginning problem solvers. Even when they begin to use the heuristic of organizing information into a table, they may find it difficult to resist the temptation of adding the common difference to the $(k - 1)$th term to find the *k*th term. Consider this table of triangular numbers:

| $k$ | $k$th △ number |
|-----|-----|
| 1 | 1 |
| 2 | 3 |
| 3 | 6 |
| 4 | 10 |
| 5 | ? |
| 6 | ? |

Most students need help in understanding that what is sought is some relationship between elements in the left-hand column and their corresponding elements in the right-hand column. One way of making this clear might be to rewrite certain elements in the right-hand column (e.g., for $k = 1, 3, 5, 7, 9$, etc.) to help the students see that other patterns are forming (see fig. 4). This technique not only helps show the way to solving the immediate problem but provides a strategy that could be used in the future.

| $k$ | $k$th △ number | Another representation |
|-----|-----|-----|
| 1 | 1 | $1 \cdot 1$ |
| 2 | 3 | |
| 3 | 6 | $3 \cdot 2$ |
| 4 | 10 | |
| 5 | 15 | $5 \cdot 3$ |
| 6 | 21 | |
| 7 | ? | $7 \cdot 4$? |

Fig. 4

At this point a reasonable conjecture might be made for the even $k$'s also (fig. 5).

| $k$ | $k$th $\triangle$ number | | Another representation |
|---|---|---|---|
| 1 | 1 | $\longrightarrow$ | $1 \cdot 1$ |
| 2 | 3 | $\longrightarrow$ | $2 \cdot 1\,1/2 = 2 \cdot 3/2$ |
| 3 | 6 | $\longrightarrow$ | $3 \cdot 2$ |
| 4 | 10 | $\longrightarrow$ | $4 \cdot 2\,1/2 = 4 \cdot 5/2$ |
| 5 | 15 | $\longrightarrow$ | $5 \cdot 3$ |
| 6 | ? | $\longrightarrow$ | $6 \cdot 3\,1/2\,?$ |

Fig. 5

From here it is an easy step to conjecture that

$$T_k = \frac{k(k+1)}{2},$$

which can be proved by mathematical induction or, for the less experienced problem solver, confirmed as the correct generalization, noting that proofs of such generalizations will be seen later.

Many students who are accustomed to being given a formula to apply may grow impatient with looking in such detail at what is happening—as will many teachers—but experiences that promote the understanding of mathematical relationships are a necessary ingredient in the process of learning to solve problems. It is especially important to note that the real work of finding a *general* solution to the problem came only after the required solution was found and the group engaged in some *looking back*. Too often the most valuable aspect of the problem-solving experience—discovering a nice generalization that can be used later or finding new strategies that might be helpful in future problem-solving encounters—is lost because one has stopped short. In the sample solution two kinds of information could be gained by looking back:

1. Problem-specific information (or what one can learn from the solution—the *product*):

    a) The numbers 1, 3, 6, 10, . . . are known as triangular numbers and should be recognized as such during future problem-solving experiences.

    b) The $k$th triangular number is the sum of the first $k$ positive integers.

    c) The $k$th triangular number may be represented by

    $$T_k = \frac{k(k+1)}{2}.$$

2. Steps taken in reaching a general solution (or what one can learn from the *process* of solution):

    a) "Grinding it out" can yield a correct solution but seldom an elegant or efficient one.

    b) A good strategy for problems in which some $k$th term is involved includes—

(1) starting with the simplest case;
(2) organizing information with a table;
(3) searching for a pattern;
(4) generalizing.

*Second level.* When the students reach the second level and are able to model their approaches on those they have observed, the teacher is ready to assume a less active role—that of a crutch, or as Snow (1970) calls it, a "prosthesis." At this point students may be ready to choose their own approaches to problem solutions, but the teacher must be ready to guide the search if necessary or to make suggestions if they come to a dead end. Problems must be carefully structured so that the students might fruitfully use the strategies they have observed. The following problems could be offered as a follow-up to the instructional sequence involving the triangular numbers.

1. Examine the following relationships:

    $$1^3 = 1^2 - 0^2$$
    $$2^3 = 3^2 - 1^2$$
    $$3^3 = 6^2 - 3^2$$
    $$4^3 = ?$$
    $$\vdots$$

    State an expression for $50^3$. Generalize.

2. Predict the sum:

    $$1^3 + 2^3 + 3^3 + \ldots + 50^3$$

3. How many ways can 50 poker chips be held among three poker players if each player must have at least one chip?

Each of these problems involves the triangular numbers in some way and may be solved using the strategy employed in modeling the solution to the triangular number problem. Such carefully selected sets of problems play an important role in teaching problem solving. They help the students internalize the notions of the mathematical structure of a problem, the strategies used in solving problems, and related problems. Many beginning problem solvers are frustrated by the thought that there are millions of problems "out there" to be solved. Carefully selected sets of similar problems will not only provide strategies for solution but also introduce students to the well-known technique called "reduced to a previously solved related problem."

*Third level.* When students begin to gain some facility in choosing solution paths and solving problems, they have reached the third level of development, and the teacher's role changes to that of problem provider. More and more challenging problems are provided to reinforce the usefulness of the strategies and to help the students collect a store of related problems for future use. At this point they attempt to solve the problems without any assistance. They share solutions with one another and thus see alternative solutions to the same problem. The teacher serves as a channel for the exchange of problems, suggests ideas for solution strategies, and perhaps helps students relate the new solutions to previously solved problems. As new levels of problems are approached, some return to modeling in instruction may be necessary.

Both Polya (1973) and Krutetskii (1976) recognize problem finding as a desirable goal in problem-solving activity and as evidence of a high level of analytic-synthetic mental activity. Teachers who are truly committed to fully developing their students' problem-solving potential should include plans to emphasize problem posing in their problem-solving instruction.

*Fourth level.* The fourth and final role of the teacher, corresponding to the fourth level of problem-solving competence in the student, should be that of a facilitator—facilitating the exchange of new problems posed by students. Such encouragement and positive reinforcement will aid students in formulating new problems similar to those already solved and finding others because of their motivation to view the world more mathematically, thus seeing it as a source of problems.

It is important that problem-solving instruction not stop short. All four aspects of the teacher's role must be accomplished for the instruction to be effective. In the movement from model to prosthesis to provider to facilitator, the teacher assumes progressively less responsibility for the activity; as students conversely assume more, they come closer to the goal of becoming independent and effective problem solvers.

### When to teach

In our third assumption, we noted that it usually takes time to develop proficiency in problem solving. It is not enough to set aside two weeks for problem solving or to include isolated problem-solving experiences on rainy days or Mondays. Carefully planned and—in the beginning—structured problem-solving instruction should be an integral part of the curriculum. In the beginning, one day a week or parts of more than one day could be set aside for problem-solving instruction. A Problem of the Day and a more difficult Problem of the Week could be posted prominently on the bulletin board.

Besides these two points—regular exposure and integration into the program—a careful selection of problems to assure students of some success is very important.

Teaching for problem solving is, in the beginning, an uphill battle. Teachers meet insecurity, apathy, and frustration even in the better students. But the rewards are worth the effort.

#### REFERENCES

Kantowski, Mary G. "Processes Involved in Mathematical Problem Solving." *Journal for Research in Mathematics Education* 3 (1977):163–80.

Krutetskii, V. A. *The Psychology of Mathematical Ability in Schoolchildren.* Chicago: University of Chicago Press, 1976.

Polya, George. *How to Solve It.* 2d ed. Princeton: Princeton University Press, 1973.

Polya, George, and Jeremy Kilpatrick, eds. *The Stanford Mathematics Problem Book.* New York: Teachers College Press, 1974.

Prielipp, Bob. Problems for Solution, Problem 3454. *School Science and Mathematics* 72 (1972): 660.

Schoenfeld, Alan. "Can Heuristics Be Taught?a' In *Cognitive Process Instruction*, edited by John Lochhead, pp. 315–38. Philadelphia: Franklin Institute Press, 1979.

Snow, R. E. "Heuristic Teaching as a Prosthesis." In *A Symposium on Heuristic Teaching,* edited by R. E. Snow. Technical report no. 18. Stanford, Calif.: Stanford Center for Research and Development in Teaching, 1970.

# 20

# Measuring Problem-solving Ability

*John A. Malone*
*Graham A. Douglas*
*Barry V. Kissane*
*Roland S. Mortlock*

DEVELOPING the ability to solve nonroutine problems is important at all levels of the formal educational process—from kindergarten through graduate school. Associated with this objective are the tasks of determining the level of problem-solving ability in individual students and determining the effectiveness of instructional programs in developing this ability. To make these determinations, some means of measurement is required.

We are considering *nonroutine* mathematical problems only—nonroutine in the sense that a student attempting such a problem possesses neither a known answer nor a previously established (routine) procedure for finding one. (An implication is that the student has not previously attempted the problem or one very similar to it.) Such problems are distinct from exercises or problems made routine by the context in which they occur; for example, instruction that immediately precedes many "word problems" in school mathematics texts makes them routine.

To measure nonroutine mathematical problem-solving ability means to quantify it. If we can measure this ability, then we have a basis for determining whether, and by how much, an individual's ability has changed and also for determining which of several instructional programs is most effective in developing it. The data at our disposal for measuring problem-solving ability will consist of student performance in problem solving. We do not have access to the ability itself but must make inferences about the ability on the basis of the performance. This necessarily involves a model that relates performance and ability. Such a model will be described later.

What are the desirable characteristics of this measurement process? Consider the analogous process of measuring the length of an object with a ruler. There are three crucial but easily overlooked properties of this everyday process:

1. Confidence that we are, in fact, measuring length and not some other characteristic (such as weight) of the object.

2. Knowledge that the measurement of the length of a particular object does not depend on the particular ruler used. We expect to obtain roughly the same result using any of a variety of rulers that are accurately *calibrated*.

3. Knowledge that the length of a particular object does not depend on the lengths of other objects. We interpret the measurement by reference to the ruler and do not need to relate it to a set of measurements of other objects.

In this analogy, the object being measured corresponds to the person; its length, to the person's problem-solving ability; and the ruler, to the items on the ability test. These three properties, taken for granted when length is measured, are no less desirable when problem-solving ability, or indeed any mental ability, is being measured. The last two properties in particular can be assumed only if a special measurement approach is adopted. In the context of mental measurement, the second property is generally know as *item-free person measurement*, since the measurement of a person's ability is independent of the particular items used to obtain it. The third property is referred to as *person-free item calibration*, since the relative item difficulties are established independently of the abilities of the particular persons involved. Measurement procedures that have these latter two properties will be referred to as *objective*.

We might attempt to achieve the first property by providing students with nonroutine mathematical problems to solve. It is crucial that the problems be nonroutine in the sense described earlier, since otherwise we cannot state with assurance that we are measuring the right thing. A set of nonroutine problems, together with the students' attempts at solving them, will yield appropriate performance data on which to base inferences about the students' mathematical problem-solving ability.

The second property is especially important. To monitor the development of problem-solving ability, it is necessary to measure student performance more than once. Herein lies a major difficulty: it is logically impossible to reuse the same test for this purpose, since the second administration would expose the student to problems previously encountered, thus destroying their novelty. A further complication is that an initially novel problem may no longer be novel after the student is exposed to a course of instruction. The problem of determining the height of a tree, given its angle of elevation from a particular point, may be novel before the study of elementary trigonometry but routine afterward. These characteristics of the problem-solving process render the traditional approaches to repeated measuring, such as using the same test twice, infeasible.

The third property is not typically characteristic of educational measurement, although it is obviously desirable. Usually a test score can be interpreted only by reference to other test scores. For example, an eighth-grade student's score on a standardized test can be interpreted only by reference to the performance of other eighth graders on the same test. A statement that a particular score lies at the seventy-fifth percentile can be interpreted only in the context of the group of eighth-grade students from which the norm was developed. The same score that places a student at the top of the sixth grade may place the same student near the bottom of the tenth grade. But the student's problem-solving ability is the same regardless of those with whom the student is compared. As Wright (1967, p. 87) has put it, "I may be at a different ability percentile in every group I compare myself with. But I am the same 175 pounds in all of them.''

These three properties are necessary for the adequate measurement of any personal characteristic, including the ability to solve mathematical problems. The first property is generally attained when given careful attention in the test construction stage and when appropriate problems have been selected. However, the second and third properties often fail to be achieved because of the type of measurement used. Are they attainable for the measurement of problem-solving ability?

An approach formulated by the Danish mathematician Georg Rasch achieves approximations to the last two properties. Rasch's approach, which uses a particular model for inferring ability from performance, will be briefly described later. Although the approach has been demonstrably successful in a number of areas of education and psychology, we shall focus here on its applicability to measuring problem-solving ability, noting in particular how several of the previously identified difficulties inherent in this important task can be (and have been) overcome at both the research and classroom levels. Applying the Rasch approach involves six stages described later in more detail. The first four deal with compiling and calibrating a pool of problems that conform to the model's assumptions and the last two with the measurement process itself.

1. A set of problems appropriate to the background of the student population is collected. (Criteria for selecting them will be discussed later.)

2. These problems are administered to a representative sample of these students, and the response to each problem is scored.

3. A statistical test of the conformity of the responses to the assumptions of the model is applied. Problems on which the responses do not meet the criteria of statistical conformity are deleted at this stage.

4. The problems are calibrated—that is, the difficulty of each problem is established.

5. Appropriate problems are selected from the pool and administered to students whose problem-solving abilities are to be measured.

6. Responses are marked and the scores are converted into measures of ability on a common scale with the item difficulties.

## The Rasch Approach to Measurement

The Rasch approach, in the context of problem solving, postulates a simple model for the interaction between a person and a problem: the person's performance on the problem (observed in terms of the validity of his or her solution) is assumed to be a function of two factors only—(1) the ability of the person and (2) the difficulty of the problem. The measurement process incorporates these two factors algebraically into a probability model that operates on the principle that the more able the student, the greater his or her chance of success and the more difficult the problem, the smaller the chance that the student will solve it. Readers interested in the mathematics of the model are invited to refer to Rasch (1960) or Wright (1977).

Computer programs have been developed to analyze the interactions between persons of differing abilities and problems of varying difficulty and to produce output that is easily interpreted. These programs are readily accessible (e.g., MLTBIN; current cost information and ordering procedures for the user's manual and program can be obtained by writing to Dr. D. Andrich, Department of Education, University of Western Australia, Nedlands, Western Australia) and easily implemented and interpreted so that the busy teacher doesn't need to be concerned with the mathematics of the Rasch model or the details of its computer implementation.

As we mentioned earlier, the model has two principal features. The first is *person-free item calibration*. The problem-solving abilities of students in the sample are taken into account at the same time as an estimate of the difficulty of each item is made. The second feature is *item-free person measurement*. When the pool of problems conforming to the model has been compiled, student ability can then be measured by any subset of that pool in such a way that different subsets of problems (that is, different tests) give rise to statistically equivalent measures for each student.

Person-free calibration and item-free person measurement mean, in the first place, that students' abilities will be on the same scale, whatever the subset of problems from which they were estimated. This feature is extremely valuable in overcoming the difficulties mentioned earlier that are inherent in pretests and posttests of problem-solving ability. The guarantee of parallel forms is automatic with the Rasch approach.

Second, they make it possible to examine problem-solving performance using *different types* of items. For example, our early research has established that a variety of arithmetic, algebra, and geometry problems *do* fit the model, thus providing some evidence of a general mathematical problem-solving ability as distinct from different abilities specific to different areas of mathematics. The advantage here is that tests can be constructed for special purposes—that is, appropriate problems can be selected from the pool for particular situations. A test can be designed to be consistent with the stage students have reached in a course or to include as wide a variety of problems as possible.

In the third place, different students can be measured with *different numbers* of problems. This is possible because the model performs a transformation of the proportion of items correct, allowing one student to be examined by a short test and another by a longer test, with the resulting ability estimates placed on the same scale. Of course, the more problems there are on a test, the more precise will be the measurements obtained; however, this adjustment feature will be appreciated by teachers who must limit the amount of testing they include in their day-to-day programs.

Fourth, new problems can be analyzed at any time for conformity to the model and those conforming added to the pool, thus enlarging the set of problems available for use.

Fifth, it is not necessary to obtain a random sample of students to whom to administer the collection of problems being tried out for possible inclusion in the final pool. On the contrary, a representative sample is more appropriate for this person-free approach to the measurement of problem-solving ability.

Finally, each response can be scored according to how close it is to a valid solution. It is especially inappropriate to score a solution to a nonroutine problem as simply right or wrong, since varying degrees of progress toward solving the problem can be made.

These six features of the model were demonstrated in research carried out during the late 1970s with Western Australian lower secondary school students (aged about fourteen years). From an initial pool of 370 nonroutine arithmetic, algebra, and geometry problems, 71 were selected for Rasch analysis on the basis of criteria listed in the next section. Of these 71, 49 were found to conform to the model and were used to measure changes in problem-solving ability in a sample of 1000 students.

The students participated in a program that ran for the entire 1978 school year (February to December in Australia) and that involved the study of an instructional package consisting of seven problem-solving strategies coupled with extensive practice at problem solving. Three separate tests composed of problems selected from the Rasch-based problem pool were administered during the year.

Rasch approaches have also been used in the development of mathematical operations tests by the Australian Council for Educational Research (Cornish and Wines 1977) and diagnostic tests by the American Guidance Services (Woodcock 1973). In a broader context, the procedures have been applied across large educational systems—for example, by the National Foundation for Educational Research in the United Kingdom (Willmott and Fowles 1974) for national survey testing and for item banking designed to aid consistency in school-based certification. In the United States, some regional boards of education—the Portland (Oregon) Public Schools, for instance—have similarly used the Rasch approach, and in other parts of the world it is being used to construct scales for both cognitive and attitudinal traits.

On a much smaller scale, one or more classroom teachers can take advantage of the model's features and establish objective measures, secure in the knowledge that some initial effort will result in a basic pool of problems that can be expanded and used with different groups of students at different times.

## Compiling, Calibrating, and Using the Problem Pool

Like the process of test standardization, the compilation and calibration of a pool of problems is a substantial task. Consequently, we recommend that several teachers combine their efforts. As in most educational enterprises, three or four heads are better than one! A team approach should yield a larger initial collection of potentially suitable problems and a larger sample of students on which the problems can be tried and calibrated, as well as easing individual scoring loads.

Problems and ideas for generating problems can be found in the many problem books, puzzle books, journals for mathematics teachers and students, enrichment materials, mathematics methods texts, and problem-solving tests that exist nowadays, as well as in some school texts.

Besides being nonroutine (according to the guidelines set down earlier), the problems that are selected should be appropriate to the level of the students to be tested in four ways: (1) mathematical background, (2) strategies required, (3) reading level, and (4) length. There is no point in giving problems that require mathematics or strategies that are inaccessible to the students, problems that are beyond their reading level, or problems that take longer to solve than their attention span.

Once the problems are selected, they must be calibrated. This involves trying them with a representative sample of students, and in the interest of accuracy as many students as possible (at least 150) should attempt each problem. The advantages of several teachers working together should now be apparent.

As we mentioned earlier, it is appropriate to use a scoring scale that indicates varying degrees of progress toward a valid solution—for instance, the five-point scale shown in figure 1. Students are directed to record every detail of their problem-solving attempt.

| Score | Solution Stage |
|-------|----------------|
| 0 | *Noncommencement* <br> The student is unable to begin the problem or hands in work that is meaningless. |
| 1 | *Approach* <br> The student approaches the problem with meaningful work, indicating some understanding of the problem, but an early impasse is reached. |
| 2 | *Substance* <br> Sufficient detail demonstrates that the student has proceeded toward a rational solution, but major errors or misinterpretations obstruct the correct solution process. |
| 3 | *Result* <br> The problem is very nearly solved; minor errors produce an invalid final solution. |
| 4 | *Completion* <br> An appropriate method is applied to yield a valid solution. |

Fig. 1

To illustrate the use of this scale with a particular problem, four actual responses to a given problem—together with the score each response earned—are shown in figure 2. A response rating a zero score is not illustrated for obvious reasons.

With any particular set of problems, it is necessary to ensure that these categories be applied consistently to all problems as well as to different attempts at each of the individual problems.

Experience has shown us that a group of three markers scoring a test with this scale and having a solution key for each problem can reach and sustain a high degree of consistency. Both within-marker consistency (a particular marker repeatedly scoring particular solutions in the same way) and between-marker consistency (different markers scoring particular solutions in the same way) have been evident. Because the skills developed by markers are used at both the calibration and the measurement stages of the whole process, the training involved serves a double purpose. It is advisable that classroom teachers employing this scale make some effort to check the consistency with which they are using it.

Attempts will be made, obviously, to select problems of appropriate difficulty for

| | Problem | Fifteen liters of milk is supplied to a kindergarten every day in ⅓-liter packs, and all of it is consumed. If 17 of the children drink two packs each during the day and the rest drink one pack each, how many children attend the kindergarten? |
|---|---|---|
| Score | | |

**1**

17 children drink 34 packs

$3 \times 15 = 45$ packs tot.

45    34

**2**

How many Packs? = 45

17 Drink 34 ~~liters~~ packs

? Drink 45 ~~tots~~ packs

So $\frac{17}{1} \times \frac{45}{1} \times \frac{1}{\underset{2}{34}} = 22\frac{1}{2}$ children

**3**

$N$ is number of children

45 packs of milk

17 drink 34 packs

$N - 17$ drink $N - 17$ packs

$N - 17 = 11$

$N = 6$ children

**4**

Total 45 packs

17 child. $\times 2 = 34$ packs

$34 + 11 = 45$

So 11 child. drink 1 pack

$17 + 11 = \underline{28 \text{ child.}}$

Fig. 2

the students concerned, but these will not always be successful. As a guideline, we have rejected from our item pool those problems on which more than 90 percent or fewer than 10 percent of the students are successful. Although these problems may adequately conform to the model, they are of little value for the particular group of students involved. If a problem is too easy, it only provides a little information about the lower bound of their problem-solving ability. If it is too difficult, it only provides a little information about the upper bound of their problem-solving ability. By way of analogy, we would not use a set of bathroom scales to weigh an elephant. The scales can indicate only that the elephant weighs more than the maximum they can measure but nothing about how much more. For similar reasons, we would not use a road scale to weigh an ant.

After the responses are scored, the scores are submitted to computer analysis using the MLTBIN program. This program uses statistical criteria to indicate which of the problems do (and do not) conform to the Rasch model. As an indication of what the teacher might expect, we remind you of the research mentioned earlier in which seventy-one problems were tried and, of these, only forty-nine were found to conform to the model. (These forty-nine comprised approximately equal numbers of arithmetic, algebra, and geometry problems.)

The reasons for some items not conforming to the model included difficulty in applying the scoring procedures to them or a tendency for the solutions of these items to fall in the scoring categories of either 0 or 4 (with most students either unable to commence the problem or solving it perfectly). For example, scorers experienced difficulty in applying the scale to this problem:

> A point $X$ inside a triangle may be connected to a point $Y$ outside with a continuous curve that crosses each side of the figure once and only once:

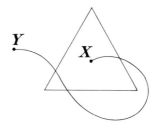

> Draw four-, five-, six-, and eight-sided figures and see if you can join $X$ to $Y$ in the same way in each. Would it be possible with a polygon of 100 sides? Why?

The 0/4 dichotomy was evident with this problem:

> What number must be added to both 100 and 164 to make each a perfect square?

This two-step problem created difficulty in applying the scale as well as producing a predominantly 0/4 score split:

> During a sale, the price of a transistor radio that had been selling for $x$ dollars was reduced by $\frac{1}{4}$ of this price. The radio was not sold, and so the sale price was reduced by $\frac{1}{3}$. What was the final sale price?

The computer program ranks the conforming problems according to their relative difficulty. This ranking is expressed on a scale that assigns larger values to more difficult problems. In the MLTBIN program, the problems are scaled to give a difficulty index that ranges from about 20 to 80. By way of illustration, sample problems and their difficulty ranking are shown in figure 3. These problems, appropriate to lower secondary school students in Western Australia, include a variety of mathematical areas and problem types.

---

### Problem 1

Complete the square so that all rows, columns and diagonals, have the same sum.

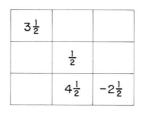

Difficulty index : 34.8

---

### Problem 2

A fireman stood on the middle rung of a ladder, directing water into a burning building. As the smoke lessened, he stepped up three rungs. A sudden flare-up forced him to go down five rungs. Later he climbed up seven rungs and worked there until the fire was out. Then he climbed the remaining six rungs to the top of the ladder and entered the building. How many rungs did the whole ladder have?    (23)

Difficulty index : 40

---

### Problem 3

In order to encourage his son in the study of algebra, a father promised to pay his son 8¢ for every problem solved correctly and to fine him 5¢ for each incorrect solution. After 26 problems neither owed anything to the other. How many problems did the boy solve correctly?    (10)

Difficulty index : 46.3

---

### Problem 4

The supermarket sends a bill for 24 dozen eggs, but leaves off the first and last digit of the cost: $_2.4_. If eggs cost less than one dollar a dozen, how much should the bill be for?    ($12.48)

Difficulty index : 47.2

---

### Problem 5

A large square $S$ has an area equal to the sum of the areas of the two smaller squares $A$ and $B$ below. To the nearest centimetre what is the length of one side of the large square $S$?    (8 cm)

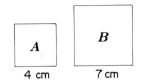

A      B

4 cm       7 cm

Difficulty index : 48.5

---

Problem 6

The minute hand of a clock is 12 cm long. An insect starts to crawl from the tip of the hand to the centre [of the clock] at a rate of 1 cm every 2½ minutes. Draw the path traced out by the insect.

Difficulty index : 56.4

---

Problem 7

In the diagram shown, determine the size of the shaded area.    (10 sq units)

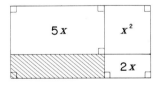

Difficulty index : 57.4

---

Problem 8

Find the total area of the right prism shown below.    (480 sq cm)

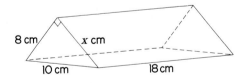

Difficulty index : 60

---

Problem 9

Each side of a triangle $ABC$ is 12 units in length. Point $D$ is the foot of the perpendicular drawn from $A$ to side $BC$. Point $E$ is the mid-point of segment $AD$. Find the length of segment $BE$.  ($\sqrt{63}$)

Difficulty index : 80.7

---

Fig. 3

Now, having compiled a pool of calibrated problems, the teacher can measure the mathematical problem-solving ability of students. At the classroom level, this task is quite straightforward. The teacher first selects a set of problems from the pool. Our experience indicates it is reasonable to expect students to attempt in a one-hour period eight to ten problems of the type illustrated. To ensure that the problems are nonroutine as far as these particular students are concerned, it is necessary to check that they have not attempted the problems previously and that they have not been taught standard methods of solving the types of problems involved. The teacher can now administer the problems, score responses according to the scale, and submit the results to the computer program. The output from the program will provide a measure of problem-solving ability for each student with a nonzero or nonperfect score. The measures so obtained are independent of the range of difficulty of the particular set of problems chosen.

The fact that student abilities and problem difficulties lie on the same scale permits a variety of conclusions to be drawn. For instance, consider the situation depicted in figure 4. The mathematics teacher can conclude from this information that—

1. Mary's problem-solving ability (65) is greater than John's (45);
2. Mary is more likely than John to solve any given problem successfully;
3. problem *X* is more difficult than problem *Y*;
4. students are more likely to solve problem *Y* successfully than problem *X*;
5. John has a better-than-even chance of solving problems of the same difficulty as problem *Y* and a less-than-even chance with problems of the same difficulty as problem *X*.

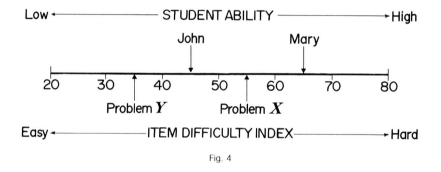

Fig. 4

## Conclusion

Determining the level of individual students' mathematical problem-solving ability and the effectiveness of instructional programs in developing this ability requires measurement. It is important that we measure the ability as well as we can. The concept of *objective* measurement, as defined in this essay, is many centuries old in the physical sciences but relatively new on the educational scene. The Rasch approach to mental measurement provides the potential for objective measurement. It deserves our close attention.

No educational gains can be made without some associated costs. It may appear to be a difficult task to compile and calibrate a pool of problems using this approach. We would contend, however, that it is no more difficult than measuring problem-solving ability with the more traditional approaches. In our view, the advantages of objective measurement outweigh the potential disadvantages.

For further information on the Rasch approach, refer to the References. For more information on the specific application of the model described here, contact the authors directly.

## REFERENCES

Cornish, Greg, and Robin Wines. *Mathematics Profile Series.* Operations Test Teachers Handbook. Hawthorn, Victoria: Australian Council for Educational Research, 1977.

Rasch, Georg. *Probabilistic Models for Some Intelligence and Attainment Tests.* Copenhagen: Danmarks Paedogogiske Institut, 1960.

Willmott, Alan S., and Diana E. Fowles. *The Objective Interpretation of Test Performance.* Berkshire: National Foundation for Educational Research Publishing Co., 1974.

Woodcock, C. *Reading Mastery Tests.* American Guidance Services, 1973.

Wright, Benjamin D. "Solving Measurement Problems with the Rasch Model." *Journal of Educational Measurement* 14 (1977):97–116.

————. "Sample-Free Test Calibration and Person Measurement." In *Proceedings of the 1967 Invitational Conference on Testing Problems,* pp. 85–101. Princeton, N.J.: Educational Testing Service, 1968.

# 21

# A New Approach to the Measurement of Problem-solving Skills

## Harold L. Schoen
## Theresa Oehmke

Mathematical problem solving involves thought processes that are rather complex. In fact, one of the more commonly used definitions of problem solving stipulates that a task must be complex if it is to be referred to as a problem (Brownell 1942). According to this definition, a task is a problem for a person if—

1. the task calls for a solution under certain specified conditions;
2. the person understands the task but does not see an immediate strategy for solution;
3. the person is motivated to search for the solution.

Problems, then, lie somewhere between *computational exercises*—for which a solution strategy is immediately known— and *puzzles*—for which there are no well-defined solution conditions that might be understood by the potential solver. It is generally agreed that developing problem-solving skills should be an important goal of mathematics instruction at all levels.

An integral part of any instruction is evaluation—determining the extent to which the learner has attained the instructional goals. Effective teachers are constantly evaluating in informal ways, but there is also a need for formal, objective evaluation. Paper-and-pencil tests of various forms are the most commonly used instruments for evaluation. Such tests can be very accurate and efficient, especially for easily measured skills. However, the complex *process* of problem solving is more difficult to evaluate. Traditionally, developers of standardized tests and most teachers have measured students' abilities in problem solving as they would simpler, learner outcomes. That is, the student is presented with a number of word problems and asked to solve them; then the number or percentage of correct answers is used as an index of the student's problem-solving ability. Some teachers refine the scoring by giving partial credit for correct procedures, but generally only the correct answers form the basis for measuring problem-solving ability or achievement. This latter procedure indicates a limited view of the complex set of skills that underlie problem solving.

Several writers and researchers have proposed multistep models of the process of problem solving. Such models suggest a number of questions about teaching and measuring problem-solving skills. If problem solving is a multistep process, would it be possible to measure a person's ability at each step? Would this be more useful information to a teacher than just the single score based on the number of correct answers? Can paper-and-pencil tests be used, or is a one-to-one interview necessary?

The answers to these questions were sought in a recent test-development effort, which is a part of the Iowa Problem Solving Project—IPSP. (IPSP is a three-year project directed by George Immerzeel of the University of Northern Iowa and funded under ISEA, Title IV, C.) The IPSP test that resulted from this effort is an alternative approach to problem-solving evaluation. Our intent here is to describe the rationale, development, and potential uses of this testing approach.

Development of the IPSP test was based on a four-step model of the problem-solving process, which is discussed in the next section. The third section discusses the reliability and validity of the test, the fourth concerns the relationship between IPSP test results and data gathered in individual interviews, and the last discusses classroom implications of the test.

## A Model for the Problem-solving Process

It is useful for purposes of instruction and sometimes for purposes of research to consider the problem-solving process as consisting of several steps. A number of different multistep models have been proposed in the literature. The number of steps ranges from three to five or more, and their names and descriptions also vary, but all the models have been proposed as attempts to refine and better define the complex process of problem solving.

The model adopted and adapted by the IPSP team is the four-step model described below.

1. *Getting to know the problem.* The problem solver is engaged in reading and discerning the meanings of the words in the problem, comprehending the problem setting, determining the relevant facts, perceiving the implied relationships, and understanding the nature of the question being asked.

2. *Choosing what to do.* The problem solver is developing or choosing a plan of attack that she or he hopes will lead to the solution. The plan could include making a table, graph, or diagram; using an equation; systematic trial and error; and so on.

3. *Doing it.* The problem solver is carrying out his or her strategy to a solution.

4. *Looking back.* The problem solver reflects on the solution in light of the conditions of the problem and attempts to answer questions such as the following: Does the solution make sense given the conditions of the problem? Can other problems like this one now be solved? Is the solution still correct if certain conditions are varied in the problem? Is there another way to solve this problem?

A heuristic model may be viewed as a guide to follow as one attempts to solve a problem. In fact, consciously or unconsciously, a successful problem solver uses at least the first three steps of this model, although obviously these are not always discrete, sequential steps toward the solution. The major goals of the IPSP were to use the model as a basis for the development of instructional problem-solving materials and a paper-and-pencil test appropriate for children in grades 5–8. The focus of this paper is the test described in the next section.

## The Iowa Problem Solving Test

If we assume that a successful problem solver employs—either implicitly or explicitly—strategies that would fall into the four categories of the IPSP model, then the problem solver has certain skills that may be similarly categorized. These skills may be considered subskills that collectively comprise an overall ability to solve mathematical problems. It would also seem reasonable that the development of these subskills should be a goal of classroom instruction and that hence a means of measuring their attainment would be useful.

Most previous attempts at studying these subskills as a part of the problem-solving process have employed an individual interview format. This is an extremely valuable technique for an in-depth assessment of problem-solving ability. However, a major disadvantage of the interview technique for classroom teachers is the excessive amount of time required to conduct interviews. This disadvantage alone is likely to limit their use as an evaluation technique to research and clinical situations (diagnosing exceptional learner characteristics, tutorial settings, etc.).

The goal of the IPSP test development was to produce an easily administered test that provides information about these problem-solving subskills that is highly correlated with data from individual interview settings. Although the IPSP test cannot provide all the information a good interviewer might gain, it does yield more valuable information than the single score from most problem-solving tests. In addition, the IPSP test appears to have the advantages of easy administration, quick scoring, and high reliability. The test provides three scores for each student: measures of that student's ability to (1) understand problems, (2) apply the solution strategies chosen, and (3) look back at the solution. After two years of effort, the IPSP team has not been able to find a valid machine-scorable way to test a student's ability to choose a problem-solving strategy, step 2 of the four-step model.

Specifically, the test contains the three subtests described below.

*Getting to know the problem (step 1)*

Items in this subtest require the student to identify extraneous or insufficient information in a problem setting, or to identify a question that could be answered using a given setting.

*Sample item:*

The school cafeteria had 230 kg of milk to be shared by 46 children. The cook wanted to know how many glasses of milk each child could have. The cook could solve the problem if he *also* knew:

1) There are 1000 grams in a kilogram.

2) Each glass holds 2 kg of milk.

3) The children all like milk.

4) Each glass is 8 cm high.

*Doing it (step 3)*

These items require a student to choose the correct computation needed to solve a problem, to compute or estimate from a diagram, or to apply a table or formula.

*Sample item:*

To convert a temperature reading from degrees Fahrenheit (F) to degrees Celsius (C), use this formula:

$$C = \tfrac{5}{9} \times (F - 32)$$

What is 59° Fahrenheit on the Celsius scale?

1) 15°          3) 27°

2) 18°          4) 74°

*Looking back (step 4)*

This subtest contains items requiring the student to identify problems that can be solved in the same way as a given problem, to determine the effect of varying the conditions in a given problem, or to evaluate a given solution strategy.

*Sample item:*

Shelley has 75 marbles, which is 11 more than twice as many as Karen has. To find how many marbles Karen has, Shelley added 75 + 11 and got 86. She then said Karen has 43 marbles. Is Shelley right?

1) Yes.

2) No. She should have multiplied 86 × 2 and got 172.

3) No. She should have subtracted 75 − 11 = 64. Then 32 is the right answer.

4) No. She should have multiplied 11 × 2 = 22. Then 75 − 22 = 53 is the right answer.

The items comprising the three subtests are interspersed throughout the test rather than arranged in blocks. In fact, the directions to the student do not indicate that there are subtests. The reason for structuring the test this way is to refrain from promoting the impression that these subskills are fragmented when, in fact, they are synthesized during the problem-solving process.

## Reliability and Validity

The IPSP test was developed over a three-year period. A group consisting of experts in educational testing and in mathematics education served as advisors in the development of the testing model and also judged whether specific items appeared to test the target skills. Items were administered to individual students in interview settings, and experimental units were developed and administered to representative samples of Iowa pupils in grades 5–8 at three different times, with revisions following each tryout.

Four forms of the IPSP test have been developed, two equivalent forms for grades

5 and 6 and two equivalent forms for grades 7 and 8. Over eight thousand Iowa children, grades 5–8, were involved in tryout testing from 1976 through 1979. Each form of the test was administered to a sample of over a thousand Iowa students in classes at each of the four grade levels.

Each test form is composed of three ten-item subtests, as described in the previous section. Students are given forty minutes to complete the test—enough time for nearly everyone to finish. Means, standard deviations, and reliability estimates for one form at each grade level are given in table 1. Results are nearly identical for the equivalent forms and are not presented here. The subtests in the table correspond to those described in the previous section. Considering that the tests are rather short, the reliability coefficients are quite high. As a point of comparison, typical reliabilities for standardized problem-solving tests, often longer than thirty items, are about .80. Of course, the reliabilities of the subtests can be improved by including more items. According to the Spearman-Brown estimate, a reliability of .70 on a ten-item test would be .82 if the number of items were increased to twenty.

TABLE 1
RELIABILITY ANALYSIS

|  | Subtest | Mean | SD | KR-8rel |
|---|---|---|---|---|
| *Unit 561* | | | | |
| Grade 5 | Step 1 | 5.41 | 2.54 | .77 |
| N = 1215 | Step 3 | 6.44 | 2.10 | .72 |
|  | Step 4 | 4.96 | 2.47 | .78 |
|  | Total | 16.81 | 6.22 | .87 |
| Grade 6 | Step 1 | 6.62 | 2.44 | .77 |
| N = 1314 | Step 3 | 7.23 | 1.87 | .68 |
|  | Step 4 | 5.99 | 2.36 | .77 |
|  | Total | 19.83 | 5.76 | .86 |
| *Unit 781* | | | | |
| Grade 7 | Step 1 | 6.16 | 2.43 | .77 |
| N = 1078 | Step 3 | 5.86 | 2.10 | .67 |
|  | Step 4 | 5.37 | 2.18 | .69 |
|  | Total | 17.38 | 5.72 | .84 |
| Grade 8 | Step 1 | 6.93 | 2.30 | .77 |
| N = 1101 | Step 3 | 6.48 | 2.08 | .68 |
|  | Step 4 | 5.96 | 2.19 | .70 |
|  | Total | 19.38 | 5.57 | .84 |

## Interviews and the IPSP Test

Data presented in the previous section indicate that the IPSP test and its subtests have high reliability. There is also evidence that the subtests appear to measure somewhat different skills. Another very important question is whether the subtests measure what they are designed to measure. In other words, is the test valid? One approach used to establish the validity of a newly developed test is to administer it to an appropriate sample concurrently with an established test that is assumed to measure the desired ability. A strong relationship between the two sets of test scores is then taken as evidence that the new test is valid.

A variation of this procedure was necessary for the IPSP test, since no machine-scorable, paper-and-pencil test exists to measure these skills. It was therefore decided that a benchmark for the validity of the IPSP test would be interview-based judgments of student performance in the three areas tested by the IPSP test. These judgments were made as students thought aloud while solving open-ended verbal problems in a one-to-one interview setting. This appeared to be the most direct way to observe and measure the three sets of subskills.

However, this decision immediately presented another problem. There is no reliable and valid technique for gathering quantitative data from individual interviews relative to the three steps in the IPSP model; although at least one recent attempt is similar to what was needed (Webb 1979). This meant that a coding and quantification system for open-ended problem-solving interviews must be developed and validated. The system had to provide measures of a student's ability to get to know a problem, to follow through with a solution strategy, and to look back once a problem is solved. The development of this measurement system proceeded over the same three-year period in parallel with the development of the IPSP test.

Initially, the moves of many students, observed while solving problems during interviews, were classified into one of the steps of the model. The actions classified in each step were then assigned a numerical value of 0, 1, 2, or 3, with 3 being the most appropriate or correct. To illustrate, the step 3 classification scheme is given in table 2.

The coding scheme underwent several revisions based on the advice of a panel of mathematics education experts and on student tryout data. The schemes for step 1 and step 4 are briefly described below.

Behavior involving reading, analyzing, and understanding the problem was classified as step-1 behavior. A score of 0 was assigned to a student who completely failed to understand a problem, a score of 1 to a student who did an incorrect analysis of the problem but understood some elements, a score of 2 to one whose analysis was correct except for a minor error like reading data incorrectly, and a score of 3 for an entirely correct understanding of the problem leading to a valid solution strategy. As in step 3, specific criteria were described for each numerical score, but a crucial point is that the step-1 score was not affected by errors in the application of a solution strategy, once chosen.

Step-4 behavior consisted of student moves after a tentative solution was reached. Many students stopped as soon as they had an answer and were given a score of 0 for step 4. Briefly, a score of 1 was assigned if some uncertainty was expressed but no systematic check was made, a score of 2 was assigned if a check was attempted but was either incorrect or incomplete, and a score of 3 was assigned if a valid check of the computation, conditions, or reasonableness of the solution was carried out. Again, specific criteria were described for each numerical score, but an important point is that the step-4 score was not affected by any behavior preceding a tentative solution. An exception was that students were assigned 0 on step 4 if no tentative solution was reached.

The following examples illustrate the coding scheme and its relationship to the IPSP test. Ann, a sixth grader, was presented with this problem:

TABLE 2
Coding Scheme for Step 3

| 0 | 1 | 2 | 3 |
|---|---|---|---|
| Any manipulations or computations the student does are incorrect. | Student does less than half the number of necessary computations correctly. | Student does half or more of the necessary computations correctly. | Student sets up problem correctly and carries out computations correctly. |
| Strategy is set up correctly, but student is not able to carry it out (e.g., student cannot solve an equation). | Student sets up equation but cannot solve it completely because of incorrect choice of operations. | Students sets up equation but cannot solve it completely because of computational errors. | Student sets up problem incorrectly, but all computation is done correctly. |
| Student tries to use a diagram, figure, or table but does so incorrectly. | Student uses successive approximation (systematic trial and error) and does the first step correctly but cannot carry it to the end. | Student makes a mistake in copying numbers but carries out computations correctly. | Student uses the algorithm or solves equation correctly (e.g., manipulates all parts of the equation correctly). |
| Student suggests a plan but cannot carry it out. | | Student makes an incorrect diagram, figure, or table but interprets and computes correctly. | When using successive approximations (trial and error), student uses information from previous trial correctly (i.e., derives all values correctly). |
| | | Student uses successive approximation and does the first few steps correctly but makes a computational error in the final step. | Student starts to execute plans, makes a computational mistake, but finds errors and corrects them. |
| | | Student does the computation correctly but makes a mistake in assigning units to the answer. | |

A bag of XL-50 brand marbles contains 25 marbles and costs 19¢. How much will 125 marbles cost?

Ann read the problem aloud and this conversation followed:

A: Uh . . . Oh, boy . . . hm . . .

I: What are you doing now?

A: I'm trying to figure out how I'll do this. Either add or multiply . . . O.K. I'm going to multiply 125 marbles times 19 cents. [Multiplies] It comes out $11.25. That's not right.

I: What are you trying to find?

A: I'm trying to get the right answer.

I: But what answer?

A: What I should do with 25, 19, and 125, because I know with those numbers I have to do something.

[Silence]

[Rereads the problem]

[Silence]

A: I want to see if I multiplied wrong. . . . [Remultiplies but is still stumped]

Ann exhibited a behavior that occurred frequently in student interviews. She tried to do some arithmetic operations using all the numbers in the problem. She lacked, or at least failed to use, analytic skills, essentially step-1 behavior. However, her computational skills and ability to use tables and diagrams, as illustrated in other problems, were good. On this problem, she was assigned 0 for step 1, 3 for step 3, and 0 for step 4. If she had made a computational error or misused an equation, she would have received a 0, 1, or 2 on step 3 (as described in table 2). A similar pattern emerged in her solution to other problems.

Ann's profile on the IPSP test is given in figure 1. The percentile ranks are based on an administration of the test to 1314 sixth graders in Iowa in the fall of 1978.

Dave is a fifth grader who was able to understand most of the problem settings presented to him but had difficulty carrying out his solution strategies. Here is a sample of a single-step problem:

Mr. Price earned $75 in each of 8 weeks. How much did he earn for all 8 weeks?

D: O.K. 75, 75, 75, . . . [adds eight 75s] O.K. 1, 2, 3, . . .

I: So you wrote eight 75s down, right?

D: Yes, O.K., that'd be . . . eight 5s would be 40. It'd be 0 and 4 on top. And eight 7s would be . . . O.K., let's see . . . hmm.

I: Can you put into words what you're thinking?

D: O.K., let's see . . . two 7s is 14; so it would be four 14s. [Writes them down and adds them.]

D: It'd be $5.60.

I: O.K., that's your answer then?

D: Yes.

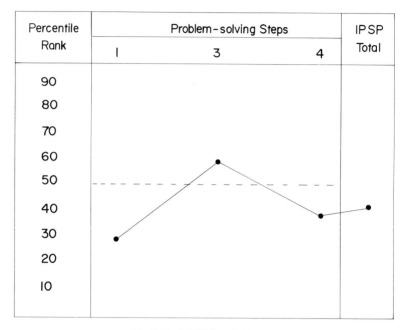

Fig. 1. Ann's IPSP Test Profile

Dave chose to add eight 75s, which was a correct strategy. However, he had diffi-culty finding the sum. To make the computation easier, he correctly noted that eight 7s is the same as four 14s. With prompting, Dave later realized that this method had left out the eight 5s, and he corrected himself. He was scored 3 on step 1 and 2 on step 3 on this problem. Dave's score on step 4 was 0, since he did not exhibit any behavior in that category.

On several other problems, Dave illustrated a fair facility for understanding and analyzing the given information. Yet he often had difficulty completing the strategy he chose. This difficulty usually arose from his choice of nonstandard, rather com-plex—though valid—strategies. His profile on the IPSP test is given in figure 2. The percentile ranks are based on an administration of the test to 1215 Iowa fifth graders in the fall of 1978.

Ann and Dave illustrate the relationship between interview findings and IPSP test results. Of course, for many students the relationship was not as clear. For many stu-dents, too, there was no comparatively strong or weak subtest score—the profile of results was nearly a horizontal line.

In order to measure the strength of the relationship between interview data gath-ered by this strategy and IPSP test scores, forty-two fifth graders in Iowa City were asked to solve five problems in an interview setting, and they also completed the IPSP test. Half the students completed the IPSP test prior to the interviews; the order was reversed for the other half. Some of the five interview problems, given below, were developed by Days, Wheatley, and Kulm (1979), and others were written by the authors.

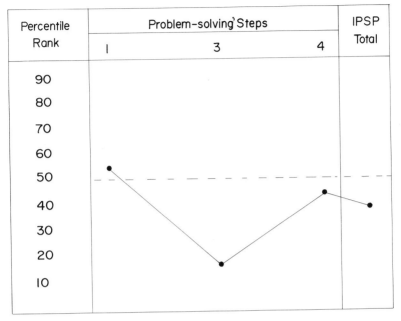

| Percentile | Problem-solving Steps | | | IPSP |
| Rank | 1 | 3 | 4 | Total |
| 90 | | | | |
| 80 | | | | |
| 70 | | | | |
| 60 | | | | |
| 50 | | | | |
| 40 | | | | |
| 30 | | | | |
| 20 | | | | |
| 10 | | | | |

Fig. 2. Dave's IPSP Test Profile

• *A rock that weighs 30 pounds on Earth weighs 5 pounds on the moon. How much does a rock that weighs 18 pounds on Earth weigh on the moon?*

• *Each hamburger at McDonald's weighs 0.15 pound before it is cooked. How much does the meat for 8 hamburgers weigh?*

• *How many pairs of socks at $1.31 a pair can you buy if you have $10.00? How much change will you have left?*

• *8 men and 12 women weigh 3000 pounds. The women all weigh the same. Each man weighs 210 pounds. What is the weight of one woman?*

• *You need to buy 10 grapefruit. They are sold 2 grapefruit for 25¢ or a bag of 10 grapefruit for $1.00. How much do you save by buying a bag of grapefruit?*

Agreement among raters for the interview coding scheme was 89 percent overall (93 percent on step 1, 84 percent on step 3, and 90 percent on step 4), and the inner consistency reliabilities for steps 1, 3, and 4 on the five interview problems were .78, .93, and .40, respectively, using Lord and Novick's (1968, p. 204) sample analog of the Cronbach estimate. KR-8 reliabilities for the IPSP subtest were .76, .85, and .74, respectively. Correlations between subtest scores on the five interview problems and the corresponding subtest of the IPSP test for steps 1, 3, and 4 were .64, .57, and .02, respectively.

According to these findings, there appears to be a strong, but by no means perfect, relationship between the two measurement techniques in step 1 and step 3. Step-4 behavior (looking back) was rarely observed. A complete discussion of the test validation can be found in Oehmke (1979).

## Classroom Implications

Two implications of the IPSP test development for the teaching of problem solving in the classroom will be discussed: the use of the IPSP test, or one similar to it, and the use by teachers of questions in the form of those on the IPSP test in homework assignments and classroom tests.

First, the IPSP test development effort illustrates that it is possible to construct a psychometrically sound test based on the three steps from the problem-solving model. The potential use of the IPSP test, or a similarly constructed one, seems clear-cut. With profiles of IPSP test results like those illustrated in the previous section, students or classes that are comparatively low in one of the steps could be provided with instruction in that area. The IPSP problem-solving modules are designed to facilitate such instruction (Immerzeel et al. 1977). These include questioning techniques and other classroom activities that concentrate on each step of the problem-solving process.

The second implication concerns questioning strategies in problem-solving instruction and evaluation. During the course of the test development, over one hundred students were observed individually as they solved problems. Their general lack of analytic skills became obvious. Many students viewed verbal problems as little more than algorithmic exercises—that is, they either saw what to do immediately or they gave up entirely. They were unable to, or at least were unaware that they should, think, analyze, and explore—activities that are at the heart of problem solving.

Part of the difficulty may lie in the verbal problems themselves. Students have known for a long time that simple verbal problems can often be solved by trying the most likely operations with the numbers: addition is tried if there are more than two numbers, subtraction if there are two numbers that are about the same size, and multiplication if there are two numbers where one is much smaller (or possibly division if it comes out even).

One way to discourage this sort of "nonthinking" is to ask questions calling for analytic or synthetic thought. Students should be asked to identify extraneous information in a problem setting, to vary conditions in a problem, or to write a question that could be solved using the given information. This focuses their attention on elements of a problem setting that they may not notice when problems are presented in the more standard way. It seems important, too, that these types of questions be used not only in class discussions but also in homework sets and on tests. This will emphasize to the students the importance of developing these skills. Following are two examples of problem settings and questioning sequences that could easily be included on a worksheet or a test.

*Example 1*

Tell which information is needed and which information is extraneous (not needed) to answer each question, *a* to *d*.

### Weekend Telephone Discount Rates

|               | First minute | Each additional minute |
|---------------|--------------|------------------------|
| Chicago       | $0.19        | $0.14                  |
| San Francisco | $0.20        | $0.15                  |

Last month Alice made 6 calls to Chicago totaling 42 minutes and 9 calls to San Francisco totaling 54 minutes. All the calls were made on the weekend.

a) On the average, how long was a call to Chicago?

b) What is the cost of an average call to Chicago?

c) What is the cost of an average call to San Francisco?

d) What is Alice's phone bill?

*Example 2*

A factory has two machines that make paper clips. Machine A produces 76 paper clips a minute and machine B produces 92 paper clips a minute. The total number of paper clips produced at the factory in one hour would be computed as follows:

There are 60 minutes in one hour;

76 × 60 = 4560 paper clips produced by machine A;

92 × 60 = 5520 paper clips produced by machine B;

4560 + 5520 = 10 080 paper clips produced in one hour.

a) Is the solution above correct?

b) Suppose machine A is speeded up to produce 85 paper clips a minute (instead of 76). What would be the total number of paper clips produced in one hour?

c) Suppose machine B is slowed down to produce 86 paper clips a minute (instead of 92). What would be the total number of paper clips produced in one hour?

c) Suppose machine B is slowed down to produce 86 paper clips a minute (instead of 92). What would be the total number of paper clips produced in one hour?

d) Suppose machine A starts at 9:00 A.M. but machine B is broken and doesn't start until 9:30 A.M.. What would be the total number of paper clips produced by 10:00 A.M.?

e) Suppose the paper clips produced in one hour are packaged in boxes holding 300 paper clips each. How many boxes of paper clips would be packaged?

Such questioning sequences require the student to attend to the data and relationships in the problem. Although there may sometimes be easier, "nonthinking" ways to get a final answer, these sequences require analytic and synthetic reasoning and may indeed help develop general problem-solving skills.

**REFERENCES**

Brownell, William A. "Problem Solving." In *The Psychology of Learning*, Forty-first Yearbook of the National Society for the Study of Education, pt. 2, pp. 415–42. Chicago: The Society, 1942.

Days, Harold C., Grayson H. Wheatley, and Gerald Kulm. "Problem Structure, Cognitive Level, and Problem-solving Performance." *Journal for Research in Mathematics Education* 10 (March 1979): 135–46.

Hieronymus, Albert N., and Everet F. Lindquist. *Manual for Administrators, Supervisors and Counselors: Iowa Test of Basic Skills.* Boston: Houghton Mifflin Co., 1974.

Immerzeel, George, Joan Duea, Earl Ockenga, and John Tarr. *Problem Solving Handbook.* Cedar Falls, Iowa: Iowa Problem Solving Project, 1977.

Lord, Frederic M., and Melvin R. Novick. *Statistical Theories of Mental Test Scores.* Reading, Mass.: Addison-Wesley Publishing Co., 1968.

Oehmke, Theresa. "Development and Validation of a Testing Instrument for Problem Solving Strategies of Children in Grades Five through Eight." Doctoral dissertation, University of Iowa, 1979.

Webb, Norman L. "Processes, Conceptual Knowledge, and Mathematical Problem-solving Ability." *Journal for Research in Mathematics Education* 10 (March 1979): 83–93.

# 22

# Problem Solving in Mathematics: An Annotated Bibliography

## Sarah F. Mason

THIS Bibliography is only a sampling of materials on problem solving that are available for the mathematics teacher's use. Many other references could have been included, and perhaps the bibliographies at the end of some of the references listed will guide the teacher to them. The Bibliography contains materials that are for use by, and of primary interest to, teachers. It does not contain research reports; they are too numerous to include and are adequately reviewed elsewhere.

The entries in the Bibliography are from books and journals. They are organized in the following sections: (1) bibliographies, (2) general works, (3) suggestions for teaching, (4) puzzles and recreations, (5) mathematical discussions containing material on problems, and (6) collections of problems. Sections 4, 5, and 6 list works that are useful primarily as sources of materials on problems, but most of the entries in sections 2 and 3 also contain problems that can be used in teaching. Some references are listed in more than one section.

The following are general sources that the mathematics teacher may find useful:

1. The *Mathematics Teacher* often has problem-solving ideas in its "Sharing Teaching Ideas" and "Activities" sections. Similarly, the *Arithmetic Teacher*'s "Ideas" section may provide useful material.

2. Several journals regularly give space to mathematical problems and their solutions, including the *American Mathematical Monthly*, the *MATYC Journal*, *Mathematics Magazine*, the *Mathematics Student, School Science and Mathematics, Scientific American*, and the *Two-Year College Mathematics Journal*.

3. Many publications of the National Council of Teachers of Mathematics contain ideas for teaching problem solving. In particular, several of the yearbooks should be mentioned:
   - The Seventeenth, *A Source Book of Mathematical Applications*

- The Twenty-seventh, *Enrichment Mathematics for the Grades*
- The Twenty-eighth, *Enrichment Mathematics for High School*
- The 1976 Yearbook, *Measurement in School Mathematics*
- The 1979 Yearbook, *Applications in School Mathematics*

Thanks are expressed to Thomas Butts, Kevin Gallagher, and Linda DeGuire for suggesting materials for this Bibliography.

## 1. Bibliographies

Hardgrove, Clarence E., and Herbert F. Miller. *Mathematics Library—Elementary and Junior High School*. Reston, Va.: National Council of Teachers of Mathematics, 1973.
Contains references to more than thirty books having problems appropriate for middle school students.

Schaaf, William L. *A Bibliography of Recreational Mathematics*. Vol. 1, 4th ed. Washington, D.C.: National Council of Teachers of Mathematics, 1970.
A bibliography of the literature on recreational mathematics. A helpful reference.

————. *A Bibliography of Recreational Mathematics*. Vol. 2. Washington, D.C.: National Council of Teachers of Mathematics, 1970.
A supplement to the original bibliography, bringing it up to date and filling in gaps and omissions.

————. *A Bibliography of Recreational Mathematics*. Vol. 3. Reston, Va.: National Council of Teachers of Mathematics, 1973.
Major changes from the previous two volumes include two new sections on classroom games and recreational activities and a glossary of terms related to recreational mathematics.

## 2. General Works

Cohen, Louis S., and David C. Johnson. "Some Thoughts about Problem Solving." *Arithmetic Teacher* 14 (1967): 261–62.
A discussion on the meaning of problem solving and the role of word problems in problem solving. This article is an introduction to and precedes two more articles by the same authors.

Henderson, Kenneth B., and Robert E. Pingry. "Problem Solving in Mathematics." In *The Learning of Mathematics: Its Theory and Practice*, edited by Howard F. Fehr. Twenty-first Yearbook of the National Council of Teachers of Mathematics. Washington, D.C.: The Council, 1953.
A useful reference for the teacher that discusses the theories of the problem-solving process and their implications for instructional approaches in the classroom.

Krulik, Stephen. "Problems, Problem Solving, and Strategy Games." *Mathematics Teacher* 70 (1977): 649–52.
A general discussion of the topics in the title.

Krutetskii, V. A. *The Psychology of Mathematical Abilities in Schoolchildren*. Edited by Jeremy Kilpatrick and Izaak Wirszup and translated by Joan Teller. Chicago: University of Chicago Press, 1976. (Originally published 1968)
The author carefully breaks down the problem-solving process into three stages that include gathering, processing, and retaining information. A valuable source of problems.

Polya, George. *How to Solve It*. Princeton, N.J.: Princeton University Press, 1957.
A classic and a must for mathematics teachers. Discusses a step-by-step process for successful problem solving.

————. *Mathematical Discovery*. 2 vols. New York: John Wiley & Sons, 1966.
Provides many problems for practice along with heuristics for solving them. Problems are at the high school and college levels.

————. *Mathematics and Plausible Reasoning*. 2 vols. Princeton, N.J.: Princeton University Press, 1954.
A discussion of logical reasoning with problems and solutions. Involves higher mathematics.

Pollak, Henry O. "Applications of Mathematics." In *Mathematics Education,* edited by Edward
   G. Begle. Sixty-ninth Yearbook of the National Society for the Study of Education. Chicago:
   University of Chicago Press, 1970.
   An explanation of the use of applications in mathematics and a description of the general stages involved
   in the problem-solving process.

Thompson, M. *Experiences in Problem Solving.* Reading, Mass.: Addison-Wesley Publishing
   Co., 1976.
   Presents six problem-solving activities that guide the reader through a successful attempt at solving a
   problem. From the Mathematics-Methods Program (MMP) developed by the Indiana University Mathematics Education Development Center (MEDC).

Wickelgren, Wayne A. *How to Solve Problems.* San Francisco: W. H. Freeman & Co., 1974.
   An aid to becoming a better problem solver. For the student and the teacher.

# 3. Suggestions for Teaching

*Arithmetic Teacher* 25 (2), November 1977.
   The entire issue is devoted to problem solving. Most articles are on some aspect of teaching problem solving.

Bader, William A. "Problem Solving via Soap Bubbles." *School Science and Mathematics* 75
   (1975): 343–53.
   The scientific method is proposed as a means for approaching problem solving. This is illustrated with the
   use of soap bubbles to solve problems that are adaptations of the famous Steiner's problem.

Balk, G. D. "Application of Heuristic Methods to the Study of Mathematics at School." *Educational Studies in Mathematics* 3 (1971): 133–46.
   The usefulness of familiarity with heuristic methods is argued through ideas and examples from analogy,
   induction, limits, and continuity.

Banwell, C., K. Saunders, and D. Tahta. *Starting Points.* London: Oxford University Press,
   1972.
   A collection of problematic starting points for teaching mathematics. Has suggestions for continuing the
   lessons and descriptions of what some students have done in the situation.

Biggs, Edith E. *Mathematics for Older Children.* New York: Citation Press, 1972.
   A description of some of the problem-solving methods used in British primary schools. Contains a wide
   variety of problems.

Brown, Stephen I. "From the Golden Rectangle and Fibonacci to Pedagogy and Problem Posing." *Mathematics Teacher* 69 (1976): 180–88.
   Presents several strategies for posing problems using the golden section as a starting point.

Carmony, Lowell A. "A Minimathematical Problem: The Magic Triangles of Yates." *Mathematics Teacher* 70 (1977): 410–13.
   The author poses the magic triangle problem as a research problem for secondary mathematics students,
   gives a solution, and makes some suggestions for implementing the ideas in the classroom.

Cooney, Thomas J., Edward J. Davis, and Kenneth B. Henderson. *Dynamics of Teaching Secondary School Mathematics.* Boston: Houghton-Mifflin Co., 1975.
   Two chapters provide a rich source for remedying problem-solving difficulties, techniques of teaching
   problem solving, and problem situations for students. Relies heavily on Polya's methods.

Dienes, Zoltan P. *The Power of Mathematics.* London: Hutchinson Educational, 1964.
   Contains a chapter entitled "The Mathematical Story" in which problems and their solutions are discussed. The concern of the chapter is with facilitating mathematical memory. Examples are from algebra
   and logic.

Earp, N. Wesley. "Procedures for Teaching Reading in Mathematics." *Arithmetic Teacher* 17
   (1970): 575–79.
   Suggestions for what might be done for and by teachers to improve students' mathematical reading and
   problem-solving skills.

Engel, A. "Outline of a Problem Oriented, Computer Oriented and Applications Oriented High
   School Mathematics Course." *International Journal of Mathematics Education in Science
   and Technology* 4 (1973): 455–92.
   Contains problems that deal with dynamic processes using flowgraphs that lead to computer-programmed
   solutions. Part of the Comprehensive School Mathematics Program (CSMP).

————. "Teaching Probability in Intermediate Grades." *International Journal of Mathematics Education in Science and Technology* 2 (1971): 243–94.
Concerned with developing the intuitive background of probability. Many examples of problems.

Erickson, Robert. "The Old Integer Game." *Mathematics Teacher* 70 (1977): 140–41.
A novel way of practicing basic skills.

Greenberg, Benjamin. "That Area Problem." *Mathematics Teacher* 64 (1971): 79–80.
The solution to a specific geometric problem is given and is suggested as an aid to a plane geometry student's insight into useful techniques of proof.

Greenes, Carole E., J. Gregory, and Dale Seymour. *Successful Problem Solving Techniques.* Palo Alto, Calif.: Creative Publications, 1977.
A guide for teachers that presents an introduction to the methods of analysis for successful problem solving. Sample problems illustrate the methods.

Greenes, Carole E., Robert E. Willcutt, and Mark A. Spikell. *Problem Solving in the Mathematics Laboratory.* Boston: Prindle, Weber & Schmidt, 1972.
Develops problem-solving strategies through lessons involving manipulative materials (attribute blocks, geoboards, multibase blocks, and Cuisenaire rods).

Henney, Maribeth. "Improving Mathematics Verbal Problem-solving Ability through Reading Instructions." *Arithmetic Teacher* 18 (1971): 223–29.
The author suggests a step-by-step analysis of problems to aid students in solving problems.

Henry, Boyd. "Some Investigations for Students of Mathematics." *Mathematics Teacher* 66 (1973): 231–34.
Proposes some questions that will stimulate inquiry into other problems.

Higgins, Jon L. "A New Look at Heuristic Teaching." *Mathematics Teacher* 64 (1971): 487–95.
A discussion of the concept of heuristics teaching with suggestions for its use in teaching problem solving.

*Hints for Problem Solving.* Topics in Mathematics for Elementary School Teachers series, bk. 17. Washington, D.C.: National Council of Teachers of Mathematics, 1969.
A helpful guide for elementary teachers; contains some sample problems in different areas of mathematics.

Holt, Michael, and Zoltan P. Dienes. *Let's Play Math.* New York: Walker & Co., 1973.
Includes a discussion on the use of games in teaching mathematics. Also includes some games for children to play (with suggested age range). Basically early childhood level.

Hughes, Barnabas. *Thinking through Problems.* Palo Alto, Calif.: Creative Publications, 1976.
A manual of mathematical heuristics that guide the student through successful problem-solving techniques. Includes problems of varying difficulty.

Hunsucker, John L. "Recreational Mathematics for Teachers." *American Mathematical Monthly* 84 (1977): 56.
A description of a recreational mathematics course offered at the University of Georgia in the spring of 1975.

Kane, Robert B., Mary A. Byrne, and Mary A. Hater. *Helping Children Read Mathematics.* New York: American Book Co., 1974.
A discussion of some helpful suggestions for teaching the reading of mathematics, especially word problems. For elementary and secondary school teachers.

Kapur, J. N. "Combinatorial Analysis and School Mathematics." *Educational Studies in Mathematics* 3 (1970): 111–27.
A discussion of the need for applications in the study of mathematics, particularly the importance of combinational analysis and implication. Contains twenty combinatorial problems, with solutions for some.

Keese, Earl E. " 'The Pit and the Pendulum': Source for a Creativity Activity." *Mathematics Teacher* 68 (1975): 602–4.
An example of an activity that can be used to stimulate creativity and problem-solving activities.

Kinsella, John J. "Problem Solving." In *The Teaching of Secondary School Mathematics,* edited by Myron F. Rosskopf. Thirty-third Yearbook of the National Council of Teachers of Mathematics. Washington, D.C.: The Council, 1970.
This discussion provides teachers with an extensive collection of techniques for teaching problem solving.

Klamkin, Murray S. "On the Teaching of Mathematics So As to Be Useful." *Educational Studies in Mathematics* 1 (1968): 126–60.
Contains a section on problem solving with some problems illustrating the points of discussion. Involves calculus and algebra.

Krulik, Stephen, and Jesse A. Rudnick. *Problem Solving: Handbook for Teachers.* Boston: Allyn & Bacon, 1980.
Many activities and suggestions for teachers to use in teaching problem solving at all levels. Contains collections of nonroutine problems and strategy games.

Krulik, Stephen, and Ann M. Wilderman. "Mathematics Class + Strategy Games = Problem Solving." *School Science and Mathematics* 76 (1976): 221–25.
A discussion of problem-solving strategies, especially the use of strategy games. Three examples are given. Useful for teachers at any level.

Mann, John E. "Polygon Sequences—an Example of a Mathematical Exploration Starting with an Elementary Theorem." *Mathematics Teacher* 63 (1970): 421–28.
The author suggests letting students explore a topic on their own. This article gives an example of such an exploration.

*Mathematics Resource Project.* Vols. 1–5. Palo Alto, Calif.: Creative Publications, 1978.
Consists of five resource files drawn from the vast amount of material available for middle school mathematics. Each resource contains a teaching emphasis on problem solving and many classroom activities with emphasis on, and teaching hints for, problem solving. Developed at the University of Oregon.

Pollak, Henry O. "How Can We Teach Applications of Mathematics?" *Educational Studies in Mathematics* 2 (1969): 393–404.
A classification of problems according to application. Involves trigonometry and calculus.

Richardson, Lloyd I. "The Role of Strategies for Teaching Pupils to Solve Verbal Problems." *Arithmetic Teacher* 22 (1975): 414–21.
The author proposes that the integration of various problem-solving strategies will enhance student competence in solving verbal problems. Two strategies are offered as alternative approaches to present practice.

Smith, Seaton E., Jr., and Carl A. Backman, eds. *Games and Puzzles for Elementary and Middle School Mathematics.* Reston, Va.: National Council of Teachers of Mathematics, 1975.
A collection of readings from the *Arithmetic Teacher.* Includes different topics in mathematics.

Sobel, Max A., and Evan M. Maletsky. *Teaching Mathematics: A Sourcebook of Aids, Activities and Strategies.* Englewood Cliffs, N.J.: Prentice-Hall, 1975.
A source for problems and solutions of problems. Written for the teacher but has sources for students.

"Teaching via Problem Solving." In *Didactics and Mathematics.* Palo Alto, Calif.: Creative Publications, 1978.
One of the didactics from the Mathematics Resource Project. Discusses how to use heuristics in teaching.

Travers, Kenneth J., Leonard Pikaart, Marilyn N. Suydam, and Garth E. Runion. *Mathematics Teaching.* New York: Harper & Row, 1977.
A chapter is devoted to problem solving, discussing the characteristics of a mathematical problem and ways of teaching problem solving.

Trigg, Charles W. "What Is Recreational Mathematics?" *Mathematics Magazine* 51 (1978): 18–21.
Includes some definitions of "recreational mathematics" and paradigms of topics, people, and publications. Enjoyable article in itself.

Troutman, Andria P., and Betty P. Lichtenberg. "Problem Solving in the General Mathematics Classroom." *Mathematics Teacher* 67 (1974): 590–97.
The authors list five problem-solving steps common to various problem-solving models, seven specific abilities related to problem solving, and some activities for the classroom.

Walter, Marion I. *Boxes, Squares and Other Things.* Washington, D.C.: National Council of Teachers of Mathematics, 1970.
A description of an experience in informal geometry, based mainly on work with construction paper and milk cartons. Contains many problems related to the unit discussed; for elementary school students.

Walter, Marion I., and Stephen I. Brown. "Problem Posing and Problem Solving: An Illustration of Their Interdependence." *Mathematics Teacher* 70 (1977): 4–13.
Using a specific mathematical problem, the authors argue the notion that problem solving and problem

posing work together to give a deeper understanding of the problem-solving act. Specific strategies are suggested.

Whirl, Robert J. "Problem Solving—Solution or Technique." *Mathematics Teacher* **66** (1973): 551–53.
The author reports methods of solution that his students used to solve the problem presented in the Gross (1971) article.

Wirtz, R. *Banking on Problem Solving in Elementary School Mathematics.* Washington, D.C.: Curriculum Development Associates, 1976.
Develops higher level problem-solving skills yet requires minimal computational proficiency. Extensive comments to the teacher.

———. *Drill and Practice at the Problem Solving Level.* Washington, D.C.: Curriculum Development Associates, 1974.
Provides problem-solving activities at the manipulative, representational, and abstract levels.

Yates, Daniel S. "Magic Triangles and a Teacher's Discovery." *Arithmetic Teacher* 23 (1976): 351–54
The author discovers an extension of a magic-triangle problem that was posed in the October 1972 issue of the *Mathematics Teacher.*

Also see Henderson and Pingry (1953), Polya (1957, 1966), and Wickelgren (1974) in section 2 and Butts (1973) in section 5.

# 4. Puzzles and Recreations

Andree, Josephine P., and Richard V. Andree. *Cryptarithms.* Norman, Okla.: Mu Alpha Theta (University of Oklahoma), 1978.
Many cryptarithms with and without discussions of solutions. Contains a section on creating cryptarithms; bibliography on cryptarithms.

Andrews, W. W. *Magic Squares and Cubes.* New York: Dover Publications, 1960.
Book-length discussion of magic squares and other figures, including four-dimensional ones.

Ball, W. W. Rouse. *Mathematical Recreations and Essays.* 11th ed. New York: Macmillan Co., 1960.
An old standard. Not a book to be glossed for a problem or two; requires study but worth the effort. Covers many areas of mathematics.

Barr, Stephen. *Second Miscellany of Puzzles.* Toronto: Macmillan Co., 1969.
Over a hundred original puzzles by the author. Many deal with topology and geometry.

Beiler, Albert H. *Recreations in the Theory of Numbers.* New York: Dover Publications, 1966.
High school algebra is all that is needed to embark on this trip through the theory of numbers.

Benson, William H., and Oswald Jacoby. *New Recreations with Magic Squares.* New York: Dover Publications, 1976.
An introduction to magic squares for people with no more mathematics than high school algebra. Contains a large portion of original material.

Boyle, Pat J., and William J. Juarez. *Accent on Algebra.* Palo Alto, Calif.: Creative Publications, 1972.
A collection of problems and puzzles including crossword puzzles, cross-number puzzles, cryptarithms, and others, emphasizing vocabulary concepts and skills of elementary algebra. Pages suitable for duplication.

Brandes, Louis G. *The Math Wizard.* Portland, Maine: J. Weston Walch, Publisher, 1975.
Many problems and puzzles with answers suitable for enrichment for the high school student.

Brooke, Maxey. *Coin Games and Puzzles.* New York: Dover Publications, 1963.
Sixty puzzles, games, and stunts involving coins. Includes games from around the world and puzzles from Dudeney.

———. *150 Puzzles in Crypt-arithmetic.* New York: Dover Publications, 1963.
Detailed solutions to the first forty-six puzzles are provided to help the beginner; answers are provided for the remainder.

Brown, Gerald G., and Herbert C. Rutemiller. "Some Probability Problems concerning the Game of Bingo." *Mathematics Teacher* 66 (1973):403–6.
The game of bingo is used to illustrate the laws of probability and as a source for hypergeometric probability problems.

Burns, Marilyn. *The I Hate Mathematics! Book.* Boston: Little, Brown & Co., 1975.
Problems that students can use as enrichment and teachers can use for reference. Includes estimation, topology, probability, arithmetic, and riddles.

Caldwell, Janet. "Magic Triangles." *Mathematics Teacher* 71 (1978):39–42.
Teacher's guide and student worksheets for activities involving magic triangles.

Carroll, Lewis. *Pillow Problems and a Tangled Tale.* New York: Dover Publications, 1958.
Two books bound together. Puzzle books by the nineteenth-century mathematician who authored *Alice in Wonderland.* Involves algebra, trigonometry, and geometry.

Crouse, Richard. "*Ripley's Believe It or Not*—a Source of Motivational Incentives." *Mathematics Teacher* 67 (1974):107–9.
The author uses Ripley as a source of facts and ideas for several examples of materials that could be used as interest-generating, problem-solving situations.

Dudeney, Henry E. *Amusements in Mathematics.* New York: Dover Publications, 1970.
A collection of every type of mathematical and logical problem; 430 problems.

————. *The Canterbury Puzzles and Other Curious Problems.* New York: Dover Publications, 1958.
Puzzles and problems from the same company of pilgrims who told tales to each other on their way to Canterbury. Puzzles are of varying degrees of difficulty and deal with arithmetic, algebra, and geometry.

————. *536 Puzzles & Curious Problems.* New York: Charles Scribner's Sons, 1967.
An excellent source of puzzles that include arithmetic, algebra, geometry, combinations, and topology. Dudeney is considered England's greatest puzzlist.

Dunn, Angela, ed. *Mathematical Bafflers.* New York: McGraw-Hill Book Co., 1964.
A collection of the best puzzles from the Problematical Recreations series of Litton Industries. Also includes some puzzles created for this volume. Involves algebra, geometry, Diophantine equations, logic, probability, and number theory.

Emmet, E. R. *Brain Puzzler's Delight.* New York: Emerson Books, 1969.
A collection of problems ranging from easy to hard. Only elementary mathematical knowledge is required.

Friedland, Aaron J. *Puzzles in Math and Logic.* New York: Dover Publications, 1970.
Presents 100 mathematical problems that deal with such areas as logic, probability, number properties, and geometry. In general, problems are moderately difficult, containing challenge and surprise.

Gardner, Martin. *The Incredible Dr. Matrix.* New York: Charles Scribner's Sons, 1976.

————. *Mathematical Carnival.* New York: Alfred A. Knopf, 1975.

————. *The Mathematical Magic Show.* New York: Alfred A. Knopf, 1977.

————. *Mathematical Puzzles and Diversions.* New York: Simon & Schuster, 1959.

————. *New Mathematical Diversions from "Scientific American."* New York: Simon & Schuster, 1966.

————. *The Numerology of Dr. Matrix.* New York: Simon & Schuster, 1967.

————. *Perplexing Puzzles and Tantalizing Teasers.* New York: Simon & Schuster, 1969.

————. *The "Scientific American" Book of Mathematical Puzzles and Diversions.* New York: Simon & Schuster, 1959.

————. *The Second "Scientific American" Book of Mathematical Puzzles and Diversions.* New York: Simon & Schuster, 1961.

————. *Sixth Book of Mathematical Games from "Scientific American."* New York: Charles Scribner's Sons, 1971.

————. *The Unexpected Hanging and Other Mathematical Diversions.* New York: Simon & Schuster, 1969.
The books by Martin Gardner listed above are collections of earlier articles appearing in the "Mathematical Games" department of *Scientific American.* These books cover the entire field of recreational mathematics.

Gardner, Martin. *Mathematics, Magic and Mystery.* New York: Dover Publications, 1956.
An explanation of how mathematical tricks such as card tricks and topological tricks work.

Golomb, Solomon W. *Polyominoes.* New York: Charles Scribner's Sons, 1965.
The concept of dominoes is extended into "polyominoes" with discussion and problems.

Greenblatt, M. H. *Mathematical Entertainments.* New York: Thomas Y. Crowell Co., 1965.
A collection of many interesting problems on such subjects as census-taking, Euler's formula, topology, Fermat's theorem, partitioning, symmetry, probability, humor in mathematics, and others.

Heafford, Philip. *The Math Entertainer.* New York: Emerson Books, 1959.
A variety of mathematical problems dealing with history, symbols, geometry, measurement, and permutations; various levels of difficulty.

Heath, Royal V. *Mathemagic.* New York: Dover Publications, 1953.
Reprint of an old favorite. Magic, puzzles, and games with numbers.

Hunter, J. A. H. *Figurets: More Fun with Figures.* New York: Oxford University Press, 1958.
Includes 150 problems dealing with numbers. Involves number theory and algebra.

—————. *Fun with Figures.* Toronto: Oxford University Press, 1956.
Gives 150 problems dealing with numbers. Includes suggested solutions for some, answers for all.

—————. *Mathematical Brain-Teasers.* New York: Dover Publications, 1976.

—————. *More Fun with Figures.* New York: Dover Publications, 1966.
Two more collections of tantalizing brainteasers by the same author. Many are in cleverly worded story form.

Hunter, J. A. H., and Joseph S. Madachy. *Mathematical Diversions.* New York: Dover Publications, 1975.
Selected topics from all areas of recreational mathematics.

Hurley, James F. *Litton's Problematical Recreations.* New York: Van Nostrand Reinhold Co., 1971.
Nearly 600 problems that first appeared in "Problematical Recreations," a series of mathematical puzzlers in trade publications, sponsored by Litton Industries. Includes logic, probability, algebra, geometry, Diophantine equations, number theory, and calculus.

Jacoby, Oswald, and William H. Benson. *Mathematics for Pleasure.* New York: McGraw-Hill, 1962.
A collection of problems that are interesting and challenging. Very few require advanced mathematical knowledge. Includes topics such as probability, whole numbers, and fun with letters.

Judd, Wallace. *Games, Tricks and Puzzles.* Menlo Park, Calif.: Dymax, 1974.
Motivational material for students of all ages. Includes discussion and pictures of inside parts of the calculator.

Kahan, S. *Have Some Sums to Solve—the Compleat Alphametics Books.* Farmingdale, N.Y.: Baywood, 1978.
An introduction to alphametics (cryptarithms).

Kordemsky, Boris A. *The Moscow Puzzles.* New York: Charles Scribner's Sons, 1972.
Presents 359 mathematical problems of varying types and difficulty. Some are similar to problems found in the works of Dudeney and Loyd.

Kraitchik, Maurice. *Mathematical Recreations.* New York: Dover Publications, 1942.
Gives 250 puzzles and problems ranging from ancient Greek and Roman problems to problems in probability, magic squares, geometry, topology, and chess.

Lake, A. *The Puzzle Book.* New York: Hart Publishing Co., 1976.
Many new and delightful puzzles.

Leeflang, K. W. H. *Domino Games and Domino Puzzles.* New York: St. Martin's Press, 1975.
*The* book on dominoes.

Lindgren, Harry. *Recreational Problems in Geometric Dissections and How to Solve Them.* New York: Dover Publications, 1972.
A unique topic of dissections of polygons. Includes mathematical areas of geometry, algebra, and trigonometry.

Loyd, Sam. *Mathematical Puzzles of Sam Loyd.* Vol. 1. Edited by Martin Gardner. New York: Dover Publications, 1959.
Puzzles, with solutions, invented by the great American puzzlist. Includes arithmetic, algebra, probability, game theory, and geometry.

—————. *Mathematical Puzzles of Sam Loyd.* Vol. 2. Edited by Martin Gardner. New York: Dover Publications, 1960.
More of Sam Loyd's puzzles.

Lysing, Henry. *Secret Writing: An Introduction to Cryptograms, Ciphers and Codes.* New York: Dover Publications, 1974.
Problem-solving strategies are used to unravel secret codes.

McConville, Robert. *The History of Board Games.* Palo Alto, Calif.: Creative Publications, 1974.
Includes a brief history of board games, variations, rules, and diagrams of the design of the game board. Also includes a section on constructing game boards.

Madachy, Joseph S. *Mathematics on Vacation.* New York: Charles Scribner's Sons, 1966.
Contains the best problems from *Recreational Mathematics Magazine*, the predecessor to the *Journal of Recreational Mathematics.* Some problems involve mathematics with practical applications. Areas of mathematics include geometry, number theory, arithmetic, and trigonometry.

Meyer, Jerome S. *Arithmetricks.* New York: Scholastic Books Services, 1965.
Contains some mathematical problems that might be useful to elementary school as well as algebra students.

Mott-Smith, Geoffrey. *Mathematical Puzzles for Beginners and Enthusiasts.* New York: Dover Publications, 1954.
Entertaining book of puzzles progressing from easy to more difficult, requiring no more than a knowledge of simple arithmetic to a knowledge of algebra and plane geometry. Involves arithmetic, logic, algebra, geometry, number theory, probability, number games, and board games.

Phillips, Hubert. *My Best Puzzles in Logic and Reasoning.* New York: Dover Publications, 1961.
Puzzles that demand clear, logical thinking. Complementary to *My Best Puzzles in Mathematics.*

————. *My Best Puzzles in Mathematics.* New York: Dover Publications, 1961.
Contains 100 puzzles that require elementary mathematics and logical thinking.

Phillips, Hubert, S. T. Shovelton, and G. S. Marshall. *Caliban's Problem Book.* New York: Dover Publications, 1961.
Contains 105 problems, mostly of intermediate or advanced level. Some require only logical reasoning; others involve mathematics ranging from geometry and algebra through permutations and number theory.

Porter, Richard D. *Project-a-Puzzle.* Reston, Va.: National Council of Teachers of Mathematics, 1978.
A collection of puzzles and problems in mathematics suitable for elementary school students. One puzzle on a page so that transparencies and duplications can be made. Each includes a solution and follow-up suggestions.

Sackson, Sid. *A Gamut of Games.* New York: Castle Books, 1969.
Many new games never before published and several lost games rediscovered.

Schuh, Fred. *The Master Book of Mathematical Recreations.* New York: Dover Publications, 1968.
Concerned with the mathematics behind puzzles, games, card tricks, and so forth. Some original puzzles and some original treatments of others.

Seymour, Dale, Mary Laycock, Verda Holmberg, Ruth Heller, and B. Larsen. *Aftermath.* 4 vols. Palo Alto, Calif.: Creative Publications, 1975.
These four books are collections of enrichment activities that involve puzzles, games, and cartoons. Many of the activities are problematic.

Shepherd, Walter. *Mazes and Labyrinths: A Book of Puzzles.* New York: Dover Publications, 1961.
Fifty puzzles along with some historical notes on mazes and labyrinths.

Silverman, David L. *Your Move.* New York: McGraw-Hill Book Co., 1971.
Mostly puzzles derived from isolated situations in various strategy games.

Simon, William. *Mathematical Magic.* New York: Charles Scribner's Sons, 1964.
Discusses different magic tricks that have a mathematical base. Includes number magic, magic squares, calendar magic, card magic, mental magic, and the magic of shape.

Smullyan, R. M. *What Is the Name of This Book?: The Riddle of Dracula and Other Logical Puzzles.* Englewood Cliffs, N.J.: Prentice-Hall, 1978.
Over 200 interesting problems dealing basically with logic and presented in an interesting and readable manner.

Spaulding, Raymond E. "Sam Loyd, America's Greatest Puzzlist." *Mathematics Teacher* 69 (1976):201–11.
A collection consumer-related puzzles of Sam Loyd, originally appearing in the *Brooklyn Daily Eagle* and the *Woman's Home Companion.*

Summers, George J. *New Puzzles in Logical Deduction.* New York: Dover Publications, 1968.
A collection of never-before-published puzzles in logic.

Wylie, Clarence R., Jr. *101 Puzzles in Thought and Logic.* New York: Dover Publications, 1957.
Good source of problems involving logic. Includes some cryptarithms.

Yawin, R. *Math Puzzles and Oddities.* Middletown, Conn.: Xerox Educational Publications, 1972.
Paper-and-pencil problems designed to appeal to students in the middle grades. Includes answers to selected problems.

## 5. Mathematical Discussions

Adler, Irving. *The Impossible in Mathematics.* Reston, Va.: National Council of Teachers of Mathematics, 1957.
A discussion of some classical problems in mathematics such as trisecting an angle, squaring the circle, and doubling the cube.

Boehm, George A. W. *The New World of Math.* New York: Dial Press, 1959.
A discussion of some "new" topics in mathematics including topology, geometry, algebra, and computer use, with ideas for problem situations for students. First appeared in *Fortune* magazine.

Bold, Benjamin. *Famous Problems of Mathematics.* New York: Van Nostrand Reinhold Co., 1969.
A discussion of the history of constructions with a straitedge and compasses and why it took so long to answer some of the classic problems. Problems are introduced throughout to complement the discussion.

Bowers, Henry, and Joan E. Bowers. *Arithmetical Excursions: An Enrichment of Elementary Mathematics.* New York: Dover Publications, 1961.
A discussion of elementary mathematical topics with problems at the end of each chapter. Appropriate for gifted elementary students as well as curious adults.

Bradley, A. Day. "The Three-Point Problem." *Mathematics Teacher* 65 (1972): 703–6.
A discussion of several historical solutions to the three-point problem.

Bright, George W. "Learning to Count in Geometry." *Mathematics Teacher* 70 (1977): 15–19.
Some specific problems that involve counting geometric regions.

Brown, Stephen I. *Some "Prime" Comparisons.* Reston, Va.: National Council of Teachers of Mathematics, 1978.
A discussion of prime numbers; includes many problems. Appropriate for better high school students.

Butts, Thomas. *Problem Solving in Mathematics.* Glenview, Ill.: Scott, Foresman & Co., 1973.
A collection of problems dealing with arithmetic and elementary number theory; includes some discussion of problem solving from both the student's and the teacher's point of view.

Conway, John Horton. *On Numbers and Games.* New York: Oxford University Press, 1976.
This book requires some effort. It is beyond the abilities of all but the most talented high school students.

Courant, Richard, and Herbert Robbins. *What Is Mathematics?* New York: Oxford University Press, 1941.
A general text in mathematics covering number theory, geometrics, topology, analysis, and calculus. A source for problems in these areas.

Dörrie, Heinrich. *100 Great Problems of Elementary Mathematics: Their Mastery and Solution.* New York: Dover Publications, 1965.
A source for problems for advanced high school students or for college students.

Engel, A. "Geometrical Activities for the Upper Elementary School." *Educational Studies in Mathematics* 3 (1971): 353–94.
A selection of problem-solving activities for grades 5–7 involving less familiar topics in geometry.

Eves, Howard *A Survey of Geometry.* 2 vols. Boston: Allyn & Bacon, 1965.
A textbook with a large number of problems in each section. Suggestions for solutions are given at the end.

Experiences in Mathematical Discovery. Washington, D.C.: National Council of Teachers of Mathematics, 1966–1971.
A series of nine booklets that use the discovery approach for such topics as geometry, arrangements and selections, mathematical thinking, and rational numbers.

Gardner, Martin. *Aha! Insight.* New York: Scientific American, 1978.
Problems that seem difficult but an aha! reaction could lead to an immediate solution.

Garvin, Alfred D. "A New Slant on the 'Biggest Box' Maximum-Minimum Problems." *School Science and Mathematics* 77 (1977): 545–50.
Gives a slanted-sides solution to the biggest-box problem in differential calculus.

Golovina, L. I., and I. M. Yaglom. *Induction in Geometry.* Boston: D. C. Heath & Co., 1963.
Problems dealing with induction are at the end of each chapter; for high school or college students.

Gross, Herbert I. "A Problem, a Solution, and Some Commentary." *Mathematics Teacher* 64 (1971): 221–24.
Presents a problem and its solution that involves logic, number theory, algebra, arithmetic, properties of number bases, aspects of magnitude, and philosophical discussion.

Hoehn, Larry. "Some Novel Consequences of the Midline Theorem." *Mathematics Teacher* 70 (1977): 250–51.
A problem is posed along with its solution, and a theorem, corollary, and construction are given that are consequences of the midline theorem.

Honsberger, Ross. *Ingenuity in Mathematics.* New York: Random House, 1970.
A collection of nineteen essays representing many different aspects of mathematical thinking. A background in high school algebra and geometry is needed for understanding. Includes number theory, geometry, combinatorics, logic, and probability. Some problems are included.

————. *Mathematical Gems.* Washington, D.C.: Mathematical Association of America, 1973.

————. *Mathematical Gems II.* Washington, D.C.: Mathematical Association of America, 1976.
Two more volumes similar to *Ingenuity in Mathematics.* These are the first two volumes in the Dolciani Mathematical Expositions (DME) series.

————. *Mathematical Morsels.* Washington, D.C.: Mathematical Association of America, 1978.
Selected problems and solutions from the problem sections of past issues of MAA publications. Some college mathematics is required.

Jacobs, Harold R. *Mathematics: A Human Endeavor.* San Francisco: W. H. Freeman & Co., 1970.
This is a general mathematics book that contains many problematic activities.

Merrill, Helen A. *Mathematical Excursions.* New York: Dover Publications, 1933.
A discussion of topics not usually covered in elementary mathematics courses. A source of problems and ideas for problems.

Pearcy, J. F. F., and K. Lewis. *Experiments in Mathematics, Stage 1.* London: Longmans, Green & Co., 1966.
A source for laboratory work involving handling, making, and creating mathematical form and structure. Includes shapes, solids, curves, games, puzzles, and patterns.

————. *Experiments in Mathematics, Stage 2.* London: Longmans, Green & Co., 1966.
A similar volume but involving more advanced mathematics such as statistics, topology, pi, the Pythagorean theorem, and other areas of mathematics.

Peck, Lyman C. *Secret Codes, Remainder Arithmetic, and Matrices.* Washington, D.C.: National Council of Teachers of Mathematics, 1961.
An enrichment for junior and senior high school students. A discussion of the topics along with problems to solve.

Prielipp, Robert W. "A Problem Involving the Division Algorithm for Integers." *School Science and Mathematics* 77 (1977): 566.
Proposes a nontrivial problem involving the division algorithm.

————. "Are Triangles That Have the Same Area and the Same Perimeter Congruent?" *Mathematics Teacher* 67 (1974): 157–59.
A proof to answer the title question.

Rademacher, Hans, and Otto Toeplitz. *The Enjoyment of Mathematics*. Princeton, N.J.: Princeton University Press, 1957.
  The authors emphasize types of mathematical phenomena, methods of proposing problems, and methods of solving problems; they deal particularly with number theory, geometry, and topology. Involves the use of arithmetic, algebra, and geometry.

Rapaport, Anatol. *Two-Person Game Theory—the Essential Ideas*. Ann Arbor, Mich.: University of Michigan Press, 1966.
  An introduction to game theory presenting the essential ideas of two-person game theory to the general reader. More of a discussion than a source of problems.

Rudd, David. "A Problem in Probability." *Mathematics Teacher* 67 (1974): 180–81.
  Considers an alternative solution to a ball-and-urn problem.

Seymour, Dale, and Margaret Shedd. *Finite Differences: A Problem Solving Technique*. Palo Alto, Calif.: Creative Publications, 1973.
  A graduated series of pattern-finding problems. Presents a systematic approach for solving them.

Steinhaus, Hugo. *Mathematical Snapshots*. New York: Oxford University Press, 1969.
  Ideas using practical applications for investigations in geometry, topology, and number theory.

*String Figures and Other Monographs*. New York: Cheslsea Publishing Co., 1960.
  Includes "Methods and Theories for the Solution of Problems of Geometrical Construction" by Julius Petersen. Discussion of, and problems involving, geometrical constructions.

Tietze, Heinrich. *Famous Problems of Mathematics*. New York: Graylock Press, 1965.
  A detailed discussion of such classics as the trisection of an angle, the four-color problem, and Fermat's last theorem.

Usiskin, Zalman. "Six Nontrivial Equivalent Problems." *Mathematics Teacher* 61 (1968): 388–90.
  Six nontrivial problems that can be solved using the same arithmetic. Appropriate for second-year algebra students.

Vilenkin, N. Y. *Combinatorics*. Translated by A. Shenitzer and S. Shenitzer. New York: Academic Press, 1971.
  An introduction to combinatorics containing over four hundred problems with solutions.

Webb, Leland F. "Variations on a Theme of the Rational Numbers." *School Science and Mathematics* 76 (1976): 401–7.
  Some problems with solutions that are within the domain of rational and irrational numbers. They basically involve analytic geometry.

Williams, John D. *The Compleat Strategyst*. New York: McGraw-Hill Book Co., 1966.
  A readable, elementary introduction to game theory. Includes many problem examples.

Yaglom, I. M. *Geometric Transformations*. Translated by Allen Shields. New York: Random House, 1962.
  Written for those already familiar with transformational geometry. Problems with solutions are at the end of each chapter.

# 6. Collections of Problems

Bryant, Steven J., G. E. Graham, and K. G. Wiley. *Nonroutine Problems in Algebra, Geometry and Trigonometry*. New York: McGraw-Hill Book Co., 1965.
  A source of 134 problems with solutions given and an additional 25 problems with no solutions given. For grades 7–12.

Burkill, J. C., and H. M. Cundy. *Mathematical Scholarship Problems*. Cambridge: At the University Press, 1962.
  Problems primarily intended for candidates for Cambridge University scholarships in mathematics, algebra, geometry, trigonometry, calculus, and mechanics; includes hints and solutions.

Charosh, Mannis, ed. *Mathematical Challenges*. Washington, D.C.: National Council of Teachers of Mathematics, 1965.
  A collection of varying types of problems that appeared in the *Mathematics Student Journal*. For grades 7–12.

Eves, Howard, and E. P. Starke, eds. "The Otto Dunkel Memorial Problem Book." *American Mathematical Monthly* 64 (7), pt. 2, 1957.
A collection of the better problems that have appeared over the years in the *American Mathematical Monthly*. Also contains a classification of problems that have appeared in the *Monthly* from 1918 to 1950.

Graham, Lloyd A. *Ingenious Mathematical Problems and Methods.* New York: Dover Publications, 1959.
Presents 100 problems involving high school mathematics, number theory, statistics, geometry, networks, and inversion. Most of the book focuses on solutions. A majority of the problems are original.

————. *The Surprise Attack in Mathematical Problems.* New York: Dover Publications, 1968.
Fifty-two problems in the same style as *Ingenious* problems.

Greenes, Carole E., Rika Spungin, and Justine M. Dombrowski. *Problem-matics: Mathematical Challenge Problems with Solution Strategies.* Palo Alto, Calif.: Creative Publications, 1977.
A collection of twenty-five mathematical problems with solution strategies. Problems from arithmetic, geometry, algebra, number theory, and logic. For students in grades 6–12.

Hill, Thomas J., ed. *Mathematical Challenges II plus Six.* Reston, Va.: National Council of Teachers of Mathematics, 1974.
A companion to *Mathematical Challenges* (see Charosh [1965]).

Kürschák, J. *Hungarian Problem Book I, II.* Translated by Elvira Rapaport. New York: Random House, 1963.
A translation of the Hungarian Eötvös Competition problems from 1894 to 1928. Includes a classification of problems according to area of mathematics. Involves number theory, algebra, geometry, and trigonometry.

Mosteller, Frederick. *Fifty Challenging Problems in Probability with Solutions.* Reading, Mass.: Addison-Wesley Publishing Co., 1965.
Contains fifty-six challenging probability problems. Some are easy, some are hard.

O'Brien, Thomas C. *Solve It.* Bks. 1–5. Chicago: Educational Teaching Aids, 1977.
A collection of basic problem-solving activities. Pages are reproducible. Designed for grades 4–9.

Polya, George, and Jeremy Kilpatrick. *The Stanford Mathematics Problem Book.* New York: Teachers College Press, 1974.
All twenty sets of problems from the Stanford University Competitive Mathematics Examinations for high school seniors, conducted annually from 1946 to 1965.

Posamentier, Alfred S., and Charles T. Salkind. *Challenging Problems in Algebra.* Vol. 1. New York: Macmillan Co., 1970.
Some out-of-the-ordinary algebra problems. Suitable for high school students with a knowledge of elementary algebra.

————. *Challenging Problems in Algebra.* Vol. 2. New York: Macmillan Co., 1970.
More of the same.

————. *Challenging Problems in Geometry.* 2 vols. New York: Macmillan Co., 1970.
A companion to *Challenging Problems in Algebra.*

Reynolds, T. D., W. M. Fraley, M. L. Getz, M. Helling, H. J. Katz, J. W. Kitchens, Jr., A. G. Myrick, and O. P. Stackelberg. *Space Mathematics, a Resource for Teachers.* Washington, D.C.: National Aeronautics and Space Administration, 1972.
A collection of mathematical problems related to space science, grouped according to mathematical topics: algebra, probability, trigonometry, geometry, and conic sections. Difficulty ranges from easy ninth-grade level to challenging twelfth-grade level applications.

Saint Mary's College. *Mathematics Contest Problems.* Palo Alto, Calif.: Creative Publications, 1972.
Problems from a problem-solving program in grades 7–9 and 10–12 at Saint Mary's College of the Holy Names. Involves arithmetic, algebra, and geometry. Divided into two sections—elementary and advanced.

Salkind, Charles T., comp. *The Contest Problem Book.* New York: Random House, 1961.
Problems from the annual high school mathematics contests sponsored by the MAA from 1950 to 1960.

————. *The Contest Problem Book II.* New York: Random House, 1966.
Problems from the annual high school mathematics contests sponsored by the MAA from 1961 to 1965.

Salkind, Charles T., and James M. Earl, comps. *The Contest Problem Book III.* New York: Random House, 1973.
Problems from the annual high school mathematics contests sponsored by the MAA from 1966 to 1972. Detailed solutions are provided in all three volumes.

Shklarsky, D. O., N. N. Chentzov, and I. M. Yaglom. *The USSR Olympiad Problem Book.* Translated by J. Maykovich. San Francisco: W. H. Freeman & Co., 1962.
Contains 320 unconventional problems in algebra, arithmetic, elementary number theory, and trigonometry. Used for competition by the Mathematical Olympiads in Moscow and the School Mathematical Society at Moscow State University.

Steinhaus, Hugo. *One Hundred Problems in Elementary Mathematics.* New York: Basic Books, 1964.
Problems that are interesting and different. Deals with algebra, geometry, practical situations, nonpractical situations, and games.

Straszewicz, S. *Mathematical Problems and Puzzles from the Polish Mathematical Olympiads.* Translated by J. Smólska. New York: Pergamon Press, 1965.
Problems of varying degrees of difficulty in arithmetic, algebra, geometry, and trigonometry.

Trigg, Charles W. *Mathematical Quickies.* New York: McGraw-Hill Book Co., 1967.
Contains 270 problems of varying difficulty. Emphasis is on the method of solution. The reader is challenged to devise a more elegant solution than what is given.

Yaglom, A. M., and I. M. Yaglom. *Challenging Mathematical Problems with Elementary Solutions.* 2 vols. Translated by J. McCawley, Jr. San Francisco: Holden-Day, 1964.
Well-known Russian problem book designed for mathematics students in upper high school grades and early years of college. Problems from the School Mathematics Circle at Moscow State University and from the Moscow Mathematical Olympiads.

# HOW TO SOLVE IT

## First.
You have to *understand* the problem. {

## Second.
Find the connection between the data and the unknown. You may be obliged to consider auxiliary problems if an immediate connection cannot be found. You should obtain eventually a *plan* of the solution. }

## Third. {
*Carry out* your plan.

## Fourth. {
*Examine* the solution obtained.